Half-Lives

A Guide to Nuclear Technology in Canada

nagi

‹son

OXFORD
UNIVERSITY PRESS

OXFORD
UNIVERSITY PRESS

70 Wynford Drive, Don Mills, Ontario M3C 1J9
www.oupcanada.com

Oxford University Press is a department of the University of Oxford.
It furthers the University's objective of excellence in research, scholarship,
and education by publishing worldwide in

Oxford New York

Auckland Cape Town Dar es Salaam Hong Kong Karachi
Kuala Lumpur Madrid Melbourne Mexico City Nairobi
New Delhi Shanghai Taipei Toronto

With offices in

Argentina Austria Brazil Chile Czech Republic France Greece
Guatemala Hungary Italy Japan Poland Portugal Singapore
South Korea Switzerland Thailand Turkey Ukraine Vietnam

Oxford is a trade mark of Oxford University Press
in the UK and in certain other countries

Published in Canada
by Oxford University Press

First Published 2009

Library and Archives Canada Cataloguing in Publication

Tammemagi, H. Y
Half-lives : the Canadian guide to nuclear technology /
Hans Tammemagi and David Jackson.—2nd ed.

First ed. published Hamilton : McMaster University
Press, 2002 under title: Unlocking the atom.

Includes bibliographical references and index.
ISBN 978-0-19-543152-0

1. Nuclear energy—Popular works. 2. Nuclear engineering—Popular works.
3. Nuclear energy—Canada—Popular works. 4. Nuclear industry—Canada—Popular works.
I. Jackson, David (David Phillip), 1940– II. Tammemagi, H. Y. Unlocking the atom. III. Title.

TK9146.T34 2009 333.792'4 C2008-907814-4

Cover image: Baris Simsek/iStockphoto

1 2 3 4 - 12 11 10 09

Oxford University Press is committed to our environment. This book is printed
on Forest Stewardship Council certified paper which contains 30% post-consumer waste.
Printed and bound in Canada.

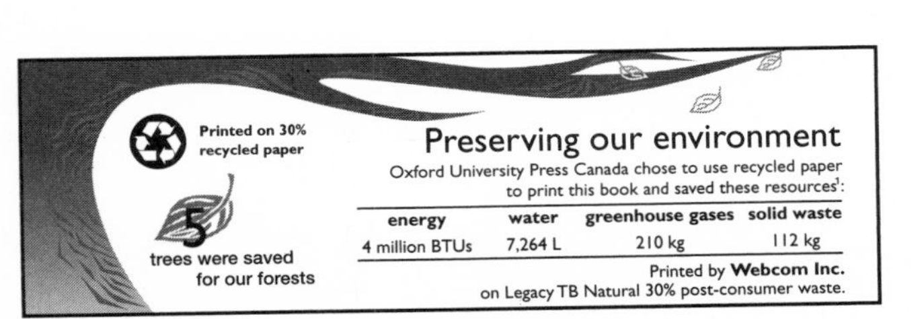

[1]Estimates were made using the Environmental Defense Paper Calculator.

Contents

Acknowledgements

We have received the generous assistance of many individuals and organizations. The patient and detailed editing of Jennie Rubio, Oxford University Press, has greatly improved this book. For their technical review, suggestions and encouragement, we would like to thank:

Tamra Benjamin, Tom Calvert, Colleen Demerchant, Dwight Foubert, Sandy Holmes, Paul Hough, Mike Krizanc , Ray Lambert, Claudia Lemieux , Dan Meneley, Alistair Miller, Yani Picard, Terry Rogers, Marie Wilson, Martyn Wash, Jeremy Whitlock.

We also thank all those who contributed to the original edition (*Unlocking the Atom*). We apologize if we have inadvertently overlooked anyone.

The support offered by the Canadian Nuclear Association, the Canadian Nuclear Society and the Organization of CANDU Industries is gratefully appreciated.

The credit for each photo or figure is indicated by the illustration. We thank the following organizations for giving us permission to use their illustrations:

Atomic Energy of Canada Limited, Best Theratronics Ltd., Bubble Technology Industries, Cameco Corporation, Canadian Nuclear Association, Canadian Nuclear Safety Commission, Canadian Centre for Magnetic Fusion, EFDA-JET, Electric Power Research Institute, Lawrence Livermore National Laboratories, Hamilton Health Sciences, International Atomic Energy Agency, Hydro Quebec, MDS Nordion, New Brunswick Power, Nray Services Inc., Ontario Power Generation, Pratt and Whitney Canada, Westinghouse.

Preface

Land's End in Cornwall, the westernmost point of England, was where I was introduced to radioactivity. What lured me to this starkly beautiful place, where waves crash against the cliffs and the barren moor rolls to the far horizon, were the rocks. My students and I spent several weeks scrambling onto high crags, climbing into quarries, and dropping down the black shafts of tin mines to retrieve samples of granite. In the evenings we pored over the day's data in pubs with ancient beams and smoke-stained walls.

As a research fellow at Imperial College, London, I was investigating the radioactivity of southwest England. Sensitive instruments at the university's research reactor site allowed me to measure the concentrations of potassium, uranium, and thorium in the rock samples. Mapping the distribution of these radioactive elements provided clues on the heat flowing from the earth and the geology of the area.

I learned that radiation has been a part of the environment since the earth was formed. In fact, energy from the radioactivity of the earth is the driving force that creates mountain ranges. The radioactivity from rocks enters soils and building materials that become a part of the homes we live in. Radioactivity is absorbed by plants through their roots, and then consumed by humans and animals as food. All rocks, soils, plants, and living creatures have always been immersed in a sea of natural radioactivity. Radiation is as natural as water or air.

Like water, a natural compound that is essential for our well being, radiation in large quantities can cause damage and must be dealt with carefully—as I learned at the reactor centre and later from the meltdowns at Three Mile Island and Chernobyl. Neither water nor radiation should be vilified for the damage they can potentially cause; rather, they need to be understood and managed with the utmost care. Radiation, of course, is far more complex than water; to many, it remains a mystery. The purpose of this book is to explain the fundamentals of radiation, nuclear energy, nuclear medicine, and other nuclear matters so they can be readily understood.

This book is unashamedly Canadian. Wherever we can, we highlight the many nuclear discoveries and inventions that Canada has produced and the often extraordinary individuals who accomplished them. Whatever one might feel about nuclear technology, Canadian achievements in this field have been significant and should be recognized as part of our national heritage. Consider the following:

- Canada was the second country to build an operating nuclear reactor.
- Canada has developed its own unique electricity-producing nuclear reactor, the CANDU, which is competitive in the world market.
- Canadians are among the world's largest consumers of nuclear electricity and have amongst the highest number of nuclear plants per capita of any nation in the world.
- Canada is the world's largest uranium exporter.
- Canada is the world's largest producer and exporter of radioactive materials for medical diagnosis and treatment. About 12 million medical procedures are conducted each year around the world using medical isotopes made in Canada.
- Canada is a world leader in the manufacture of radiation devices for medicine and industry including cancer therapy machines, food irradiators, sterilizers for medical products, and irradiators for industrial processes.

Although the construction of new nuclear power plants in North America laid dormant for a long period, the situation has changed dramatically since the first edition of this book. Concerns about global warming and rapidly diminishing supplies of oil and gas have caused a renewed interest in nuclear electricity. Canada is on the verge of building new reactors in Ontario, New Brunswick, and possibly in Alberta and Saskatchewan. With the world teetering on the edge of energy and environmental crises, it was time to bring out an updated edition of this book.

Of course, the reasons for the original book are still valid. Nuclear technology and radioactivity remain highly controversial subjects. Nuclear power, like cloning and genetically modified foods, attracts suspicion and antagonism. Opposition is routinely mounted against all proposed nuclear projects whether they are power plants, waste disposal facilities, or food irradiation centres. For many, nuclear technology is a symbol of the problems of modern society such as a deteriorating environment and overly-powerful mega-corporations.

On the other hand, the nuclear industry, and sometimes governments, view this opposition as an emotional reaction, rooted in irrational fears. Since the nuclear scientists and engineers are the experts, they adopt a "we know best" attitude in the face of opposition.

A common feature in the nuclear debate is the hurling of charges and countercharges, many of dubious validity. For the nuclear debate to move forward, communication must be based on facts. This book is intended to help remedy that situation and turn the debate into a dialogue.

Using plain language, this book explains the fundamentals of nuclear reactions and radioactivity and how the various nuclear technologies work. All facets of nuclear technology are explained, from uranium mining to electricity-producing nuclear power reactors—including the Canadian-designed CANDU reactor—to nuclear medicine and industrial applications. The history of nuclear technology in Canada is described, as are the health impacts of radiation and the technical controversy that surrounds this topic.

You will also be able to assess nuclear power and wastes in relation to their alternatives.

This book is intended for those who wish to learn about nuclear science and technology in the Canadian context. A few chapters will require some high-school science and a bit of mathematics, but the main prerequisite is an interest in learning about the subject. Appendix A contains an explanation of radiation and nuclear processes. New words and acronyms are bolded the first time they appear in the text and are explained in the glossary. There are many figures and photographs. Because it is relatively comprehensive, this book is a useful reference in Canadian universities, colleges, and high schools, as well as a resource for anyone who wishes to gain a better understanding of nuclear technology.

Hans Tammemagi, 2009

-1-

Nuclear Technology at the Crossroads

The period from 1950 to 2000 will be remembered as the Golden Era of modern civilization, the pinnacle reached by humans after a million years of evolution. This brilliant half-century was sponsored largely by fossil fuels, especially oil, which brought unprecedented economic growth, plentiful transportation, and a rich and diverse lifestyle.

But in the new millennium the shine has worn off, and civilization is suddenly at one of the most important turning points in history.

The oil bonanza allowed human population to grow beyond the carrying capacity of the planet. Today, humans are teeming on the globe. In 2008, the earth's population reached 6.7 billion—an astonishing seven-fold increase in the past two centuries. And growth continues. The United Nations calculates that by the year 2050 the population could be 10 billion people!

A pressing problem is the arrival of "peak oil." There is no better signal than the price of oil, which in mid-2008 skyrocketed past $130. The price remains both volatile and unpredictable. Peak oil, this two-syllable piece of jargon, is another way of saying that from now on the supply of oil will diminish each year, even as population and demand continue to grow. This is a sobering thought: our modern industrial society is totally dependent on this versatile fuel. It is the foundation for transportation, industry, agriculture, fishing, and more.

In addition, the world faces serious food shortages. It may have taken two centuries, but the Malthusian Devil is finally banging on the door. For seven of the eight years to 2007, global production of cereal grains has not met consumption. The price of cereal crops such as rice, corn, and wheat doubled in the period from mid-2007 to mid-2008. Poor nations are hardest hit and food riots have erupted in over ten countries.

The United Nations announced in 2008 that large segments of the world face immediate hunger, and global food production must be doubled in the next 30 years.

But how is this possible? There are no empty lands to cultivate and agriculture is highly dependent on oil and gas to power machinery, to make pesticides and fertilizers, and for shipping. Food prices are rising in lock-step with the price of oil. And now another complication: the rush to harvest cereal grains like corn to make biofuels for cars rather than food for people.

The world's food situation is grim, and it can only get worse, given that we are adding some 70 million more people to the planet every year.

And it does not stop with the oil and food crises. We're also staring down the throat of global warming—probably the most insidious threat ever faced by humans. Global warming, caused largely by burning massive quantities of fossil fuels, will lead to dramatic weather changes that include floods in low-lying coastal areas and drought in the prairie bread basket.

Overpopulation, peak oil, food and water shortages, and global warming are the ingredients for a perfect storm. How do we avoid it?

First, we must abandon the fantasy that renewable energy sources like wind and solar will save the globe. Although clean, energy derived from renewable sources simply can't be supplied in enough quantity to fill the enormous demand. And coal, although abundant, sends pollution and copious quantities of carbon dioxide into the atmosphere; it will only make the problem worse. The world desperately needs clean, affordable energy, and lots of it.

Clearly, we need to implement more conservation and develop a smaller eco-footprint with hybrid cars, smaller homes, diets with less meat, more bicycling, and better recycling. But with an ever-growing population, these initiatives will only buy a little time.

Short of a dramatic decrease in population, we have no choice but to use advanced technologies—and continue to develop new ones.

Nuclear technology is a powerful and complex tool that has enormous potential to help in these difficult times. If properly harnessed, it can provide an almost limitless supply of electricity. Nuclear power can also be used to make hydrogen, which could replace oil as a transportation fuel. And nuclear radioisotopes and research can continue to make significant contributions to medical practice and the improvement of human health. There is, however, also a dark side, for nuclear energy has enormous destructive potential.

As we march closer to energy and food shortages, a dialogue about nuclear energy and technology is vitally important. But many people—perhaps the majority—know little about the topic aside from what they hear in the media. Many worry that a major reactor accident, such as at Chernobyl, could happen in Canada. Or that radioactivity, even in small quantities, can damage human health. It is easy to feel alarmed by the thought of an invisible material, radioactivity, which is believed to cause genetic deformities. And there is also the concern about safe disposal of nuclear wastes, seen by some as one of the most toxic substances ever produced by humans.

Similar uncertainties and fears surround other technologies such as genetic modification of food, incineration of municipal waste, and the manufacture and application of pesticides. Every tool that humans have ever invented can be misused. Pesticides and synthetic chemicals, for example, have widely contaminated the environment. On occasion—think of the 1984 industrial disaster in Bhophal, India—they have wreaked enormous devastation. But it is worth remembering that the exploitation of oil and gas has also led to many deadly explosions and widespread contamination. Think of accidents such as the sinking of tankers like the *Exxon Valdez*. Even hydroelectric power, often thought to be benign, has led to tens of thousands of deaths from dam failures. And the ubiquitous automobile claims tens of thousands of lives every year through accidents and emissions. We use these tools because they also provide significant benefits—and they are essential to support the enormous population.

For society to proceed into the new millennium we need the aid of large-scale, advanced energy sources. The critical question is this: how do we control and use nuclear—and, for that matter, all the advanced technologies available to us—in ways that bring benefits, without jeopardizing our health and the environment? How do we make appropriate decisions, and enforce them?

Making rational decisions requires an understanding of the science behind the technologies as well as thorough analysis and objective dialogue. This is not an easy task. As our society grows increasingly complex, it becomes more difficult to keep abreast. We are living in an age of information overload. Nevertheless, objective analysis must be our goal.

The purpose of this book is to present the science and engineering behind nuclear technology in an accessible format. We also place nuclear technology in perspective with other energy technologies. We try to be unbiased; we want primarily to remove the polemic and fears that surround this often loaded issue. Our objective is to provide the information, to help the reader to make up his or her own mind.

-2-

Splitting the Atom

The Dawning of the Nuclear Era

Although radioactivity plays a central role in the shaping the earth (not to mention the universe at large), we knew little about it until the late 1800s. Because radiation is invisible and emanates from the nucleus of atoms—an almost incomprehensibly tiny region that cannot be seen even with the most powerful modern microscope—it was a formidable challenge to unravel its mystery. Given the limited tools that were available in the late 1800s and early 1900s, the scientists of the time were truly ingenious.

Radiation was discovered accidentally by Wilhelm Roentgen in 1895 in Bavaria. Experimenting with a gas discharge tube in a darkened room, he noticed that a crystal some distance from the discharge tube would glow whenever the tube voltage was turned on, but not when the voltage was turned off. Something was being transmitted across the space between the tube and the crystal. Roentgen had discovered X-rays.

Around this time in 1896, Frenchman Antoine Henri Becquerel accidentally discovered that uranium ore produces an image on a photographic plate, although he could not explain why. Marie Curie (who was born in Poland) and her French husband, Pierre Curie, later determined what was happening and gave the name "radioactivity" (from the Latin, *radius*, meaning ray) to the phenomenon. In 1903, Becquerel and the Curies jointly received the Nobel Prize for their discovery of radioactivity.

Remarkable advances in the basic understanding of physics and chemistry were made in those days. Nobel Prizes, the highest recognition for scientific achievement, were awarded to many of the pioneers who so greatly improved our understanding of nuclear science.

In 1897, Ernest Rutherford, a New Zealander working at the famous Cavendish Laboratory at the University of Cambridge, England, discovered that radiation from uranium is composed of two parts, which he named "alpha" (a larger particle with a positive charge) and "beta" (a smaller particle with either a positive or a negative charge) rays.

He also astounded his colleagues by showing that alpha or beta decay could change one element to a completely different one. This ran totally counter to the understanding of the atom at that time and many scientists were skeptical; some even called it "alchemy." In 1911, he made another momentous discovery: he demonstrated that the atom has a nucleus that concentrates nearly all of the mass and all of the positive charge in a very small volume at the centre of the atom (Reeves, 2007). Hans Geiger was his colleague in this famous experiment.

Although the electron was discovered in 1900 by J.J. Thomson at the Cavendish Laboratory, it was not until 1913 that the Danish physicist Niels Bohr proposed the model of the atom as we know it today, in which the electrons surrounding the nucleus occupy distinct "energy states."

In 1932, James Chadwick at the University of Cambridge discovered the neutron, a particle with no charge which plays a central role in nuclear fission. By that time, the broad outline was in place for our present understanding of the atom and its nucleus.

Nuclear Fission

In 1905, the great physicist Albert Einstein showed that—in theory—mass and energy were equivalent. It would be more than thirty years, however, before scientists discovered the immense energy that could be released by transforming matter in the fission process. In fact, both Einstein and Rutherford despaired at times of ever finding a way to extract this energy. In 1932, a frustrated Einstein said, "There is not the slightest indication that (nuclear) energy will ever be obtainable. That would mean that the atom would have to be shattered at will."

Shortly thereafter, a Hungarian physicist, Leo Szilard, actually took out a patent on a device that would release enormous energy—a chain reaction of an atomic nucleus. This was based on a neutron-capture process involving the release of more than two neutrons. Although he had no idea whether this would work in practice, the concept described exactly the way in which a nuclear reactor works.

The next major discovery was the actual fission process itself. In 1938, Germans scientists Otto Hahn and Fritz Strassman reported a

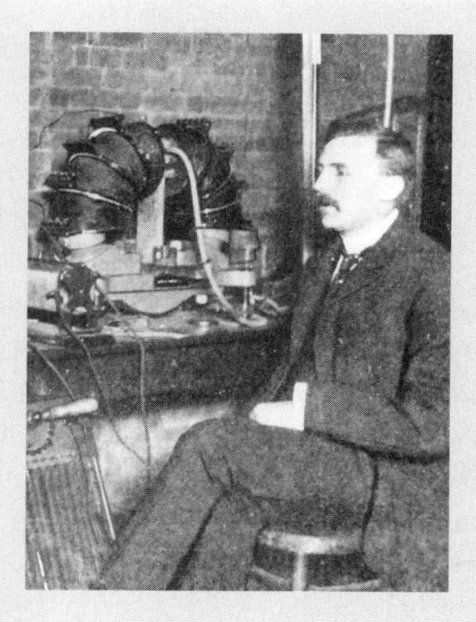

Ernest Rutherford (1871–1937), a New Zealander, is considered to be the father of nuclear physics. Although he spent most of his career in England, some of his best work was done in Canada, where he held the Chair of Physics at McGill University in Montreal from 1898 to 1907. This picture shows him in his McGill laboratory. Rutherford spent the latter part of his career at the University of Cambridge where he trained many physicists who later became prominent in Canada's nuclear development. He won the 1908 Nobel Prize for chemistry, was made a baron in 1931, and is buried in Westminster Abbey.

puzzling observation: when they bombarded uranium with neutrons, the elements barium and krypton were always produced (along with many others). Shortly after, Lise Meitner—a Jew who had escaped from Nazi persecution in Germany to Sweden—and her nephew Otto Frisch noted that barium has 56 protons and krypton has 36. The total, 92 protons, is the same as uranium. This clue led to their deduction that the uranium atom had been split into two separate elements. In other words, it had undergone a process that was to become known as fission.

But there was something even more astonishing. In the process of splitting, the uranium atom releases an enormous amount of energy. In fact, the splitting of one uranium atom releases fifty million times the energy produced by burning one atom of carbon. The potential for creating energy from fission, including its application in weapons, was immediately recognized. The observations had all come together, including Einstein's earlier theory of mass and energy equivalence.

In one typical fission reaction, a neutron strikes a uranium atom and splits it into a krypton atom and a barium atom and three neutrons, at the same time releasing a large amount of energy. Scientists express this in mathematical form, explained in more detail below, as:

$$^{235}\mathbf{U} + \textbf{neutron} \longrightarrow {}^{88}\mathbf{Kr} + {}^{145}\mathbf{Ba} + \textbf{3 neutrons} + \textbf{Energy}$$

This chemical equation, written as it is, tells us how the uranium atom splits. When an atom of uranium with atomic mass 235—235 equals the number of protons, 92, plus the number of neutrons, 143—is struck by a neutron, it changes to an atom of krypton with atomic mass of 88 (protons plus neutrons), plus an atom of barium with atomic mass of 145. In addition, there are 3 neutrons released, as well as a large amount of energy. Note that the total number of protons plus neutrons stays the same on both sides of the reaction—that is, before and after the arrow.

There are many different fission reactions for uranium-235. The left side of the equation is always the same; but on the right, there is a range of possible products of fission.Another fission reaction splits uranium into strontium and xenon. This can be written as:

$$^{235}\mathbf{U} + \textbf{neutron} \longrightarrow {}^{90}\mathbf{Sr} + {}^{144}\mathbf{Xe} + \textbf{2 neutrons} + \textbf{Energy}$$

The most probable fission products are those with atomic masses approximately 90 or approximately 140: common examples here are strontium-90, cesium-137, iodine-129, krypton-88, xenon-133, etc. In addition, a fission reaction releases either two or three neutrons (2.43 on average). These generally have high energy—or, put otherwise, they have high speed.

Finally, considerable energy is released. This is accompanied by a small loss of mass in the system. This agrees with Einstein's famous statement that mass is a very concentrated form of energy:

$$\mathbf{E = mc^2}$$

where E = Energy, m = mass, and c = speed of light.

The Fission Reactor

What is the secret to unlocking the enormous reservoir of energy locked up in the nucleus of a uranium atom? Early scientists noted that it was the neutron that caused the uranium nucleus to split (or undergo fission). They also noted that several neutrons were released during fission. If one of those neutrons could be made to hit another uranium nucleus, it could cause that nucleus to also undergo fission. Then several more neutrons would be emitted that could cause more uranium atoms to fission and so on. This is called a chain reaction, and is depicted in Figure 2-1.

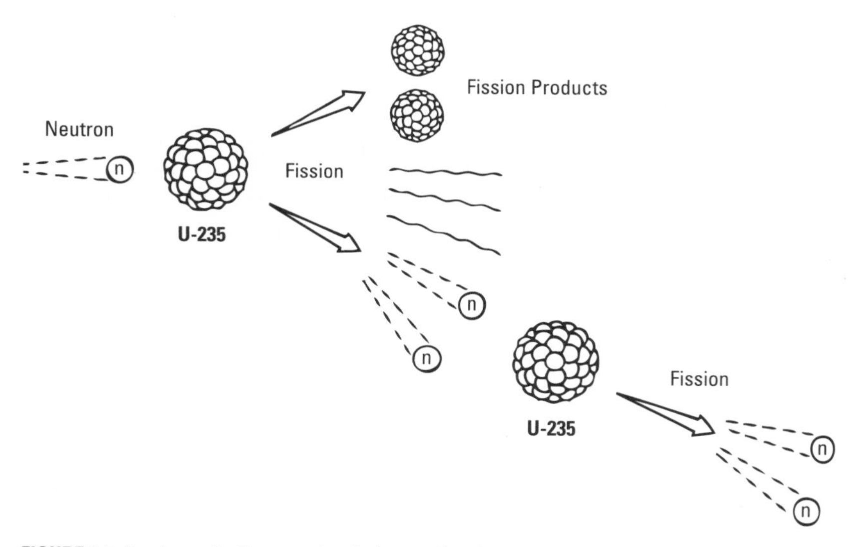

FIGURE 2-1: A schematic diagram of a chain reaction. A neutron enters from the left and causes a U-235 nucleus to fission. This fission event produces energy, in this case two more neutrons, and fission "products." The neutrons produced by the first fission then cause more fissions, which in turn produce more neutrons and the process proceeds in that manner.

If fewer neutrons are being generated by fission than are being used to initiate fission reactions, the process is not self-sustaining: it is then called "sub-critical." This is the case in nature with uranium ore bodies. If exactly the number of neutrons are being generated as are being used to split nuclei, the nuclear reaction is said to be critical. In this case a controlled amount of energy is being constantly released in a sustained chain reaction. This is the process that is used in nuclear reactors. Finally, if more neutrons are being generated than are being consumed, the chain reaction can increase extremely rapidly. This process (called "super-critical") is used in a nuclear bomb.

It is a complex matter to set up conditions leading to the process being critical or super-critical. One problem is that the nuclei of most atoms absorb neutrons and this absorption removes them from sustaining the chain reaction. So any attempt to create a chain reaction must minimize the presence of neutron absorbers. Chain reactions do not happen in uranium ore bodies, for example, because the uranium consists mostly of uranium-238—this has a low concentration of the fissionable uranium-235. The ore body also contains too many neutron-absorbing impurities. (There is a notable exception where a natural nuclear reaction occurred in a uranium ore body about two billion years ago; see Chapter 10.)

It took scientists four years to discover how to achieve the right conditions. First, there needs to be a core of fissile material—that is, material that will fission. Uranium-235 is currently the primary material used.

By a quirk of nuclear physics, the fissile atom splits most readily if the bombarding neutrons have low energy—that is, when they are going quite slowly. As neutrons emitted by the fission process have high energy, the core needs to be surrounded by a material called a moderator that slows the neutrons. Only a few materials are good at moderating neutrons without absorbing them. The more equal the nuclear mass of the moderating material is to the mass of the neutron, the more the neutron is slowed down at each collision. It was discovered that the best moderators are carbon (usually in the form of graphite) and so-called heavy water. Here, the hydrogen atom is replaced by its isotope deuterium, which has a neutron and proton in its nucleus (the nucleus of a hydrogen in normal water has only a proton, not both). Thus, deuterium is about double the weight of hydrogen, and "heavy" water is about ten percent heavier by weight than normal water. It is the deuterium in heavy water that does the moderating. Ordinary water also moderates neutrons but, because of its relatively high absorption, is not as effective as heavy water.

In general, the presence of neutron-absorbing materials should be minimized in a reactor. But at the same time, it is worth noting that these materials can also be used to stop or control the nuclear fission process. For example, neutron-absorbing control rods can be moved into and out of the core to control the reaction. Substances that are particularly good at absorbing neutrons and therefore "poisoning" the fission reaction are boron, cadmium, and gadolinium.

The engineering behind nuclear reactors is further described in Chapters 6 and 7.

The dark shadow of World War II now falls across the history of nuclear technology. Spurred by the threat of their antagonists developing a bomb, scientists and engineers on both sides of the conflict worked feverishly, compressing what would likely have taken many decades into a few years. Italian scientist Enrico Fermi showed that when neutrons are slowed down by moderators such as graphite and water they can be absorbed by nuclei, which become radioactive. He applied his theory to make several radioactive materials, for which he was awarded the Nobel Prize in 1938. He used the trip to the award ceremony in Stockholm to escape from fascist Italy, and later played a key role in the US nuclear program.

On the eve of World War II, Einstein and Szilard wrote to US President Franklin D. Roosevelt, warning him that Nazi Germany might be developing the ultimate weapon of destruction. This led President Roosevelt to establish a committee to direct research on this vital topic. In 1942, shrouded in secrecy, the famous Manhattan Project was initiated.

By the fall of that year, under Fermi's direction, the world's first nuclear reactor was constructed at the University of Chicago on a racquets court under the stands of the football stadium. This primitive reactor was known as a "pile" because it consisted of a pile of graphite blocks and natural uranium, into which were inserted neutron-absorbing rods to control the reaction. The first man-made self-sustaining fission chain reaction took place on December 2, 1942. Uranium, used for adding colour to glass and ceramics for hundreds of years, now took centre stage as a powerful new energy source.

Other important nuclear laboratories were established at Oak Ridge, Tennessee, Hanford, Washington, and Los Alamos, Nevada, culminating in the development of the atomic bombs that were exploded over Hiroshima and Nagasaki, Japan. Those bombs were to have a profound effect on the outcome of the war as well as on post-war developments. Indeed, those bombs still cast an ominous shadow on modern society.

Canada Starts its Nuclear Program

The nuclear era in Canada began in the midst of the frantic race between fascist Germany and the Allies to prepare an atom bomb. The Canadian team included French and British scientists who had been evacuated from Europe. Sir John Cockcroft, who later became director of the Canadian effort, was one of the more eminent of these scientists. Cockcroft was a student of Rutherford and his research from the Cavendish Laboratory in Cambridge, England was transferred to Canada.

Those were drama-filled years. Two scientists, Hans Von Halban and Lew Kowarski, escaped from the Institut du Radium in Paris only steps ahead of the invading German army. They took with them 200 kilograms of heavy water, which was difficult to manufacture—the world's total supply. Both the scientists and the heavy water became key parts of the embryonic nuclear research effort in Canada (Bothwell, 1988).

Canada's program was designed to support the larger American project by developing a heavy-water reactor to produce plutonium, a prime fissile ingredient for the atomic bomb. Initially located in Montreal, in 1944 the project team moved to specially constructed laboratories at Chalk River, Ontario, 200 kilometres northwest of Ottawa on the Ottawa River. The research program was directed by the National Research Council, which was already involved in nuclear research under the direction of George Laurence.

Uranium, up to then an unwanted by-product of the Eldorado radium refinery at Port Hope, suddenly became a very valuable commodity—as did the Eldorado refinery, the only radium refinery in North America.

The objective of Canada's nuclear program was to construct a reactor moderated by heavy water with thermal power output of 20 to 40 megawatts (million watts of energy). To better understand the physics and material problems, a small prototype reactor named Zero Energy Experimental Pile (ZEEP) was designed and went critical on September 5, 1945. This was a remarkable achievement, as Canada became only the second country in the world, after the United States, to control nuclear fission using a reactor. In 1966, ZEEP was named an Ontario historic site and in 1970 it was shut down. Its calandria, the steel vessel inside which the uranium fissions, is now on display at the Museum of Science and Technology in Ottawa.

Nuclear development in Canada was initially the responsibility of the National Research Council. In October 1946, the Atomic Energy Control Board was created to be the regulatory body, independent from the developer. The Atomic Energy Control Board (AECB) was renamed the Canadian Nuclear Safety Commission (CNSC) in 2000. (To avoid confusion, the CNSC will be used even for events that occurred when it was still officially called the AECB.)

From the outset the goal was to develop peaceful applications of nuclear technology. Canada has an exemplary track record in this regard. Canada is one of the world's leading countries in nuclear technology. Since 1946, this has included the capability to develop nuclear weapons; but Canada has chosen not to do so. In fact, a weapons

Thermal and Electrical Energy

The power of reactors is measured in the unit megawatt (MW) or one million watts. For research reactors, which do not generate electricity, the power is quoted in thermal megawatts. For example, the McMaster University Reactor in Hamilton, Ontario is rated at 5 megawatts (thermal).

For a power reactor, electricity—not heat—is the quantity of interest. Even with modern equipment, only about 33 percent of the heat created by a nuclear reactor is converted to electricity. Thus, for a CANDU reactor of the type located at Pickering, capacity is quoted as 515 megawatts: that is, it can generate 515 megawatts of electricity (even though it generates about 1550 megawatts of thermal power). This can also be written as 515 megawatts (electrical) or 515 MWe.

In this book, "megawatt (thermal)" or MWt always means thermal megawatts, and "megawatt" always means electrical megawatts.

program for Canada has never been seriously considered. All Canadians should take pride in this fact.

Following World War II, Canada continued to support the US and British weapons programs as a member of the NATO alliance, and continued its efforts to develop a powerful research reactor. In 1947, the 20-megawatt thermal National Reactor Experimental (NRX) reactor came online at Chalk River. Although its original purpose was to supply weapons-grade plutonium for the United States, mechanisms for producing isotopes and channels for conducting experiments with beams of neutrons were also built into the reactor. NRX generated a very high neutron flux (number of neutrons per second per unit area) and for many years was the world's most powerful research reactor, while at the same time producing plutonium. In 1965, under Lester Pearson's government, Canada made the policy decision to cease contributing nuclear materials to the US weapons program.

Following an accident in 1952 (see Chapter 8), NRX was rebuilt and upgraded to 40 megawatts. NRX had a long history of producing radioactive elements for medical applications that began in 1949 with sales of iodine-131, phosphorus-32, carbon-14, cobalt-60, and others. NRX has also had an illustrious career in research that is discussed in Chapter 16. NRX was finally taken out of service in 1992, at the age of 45—this is remarkable given that its useful life was originally estimated to be about five years.

In 1952, Atomic Energy of Canada Limited (AECL), a Crown Corporation, was formed and took over responsibility for Canada's nuclear program from the National Research Council when it became evident that the work was beginning to involve industrial-scale activities.

In 1957, a larger reactor, the National Research Universal (NRU), started up in Chalk River. Now rated at 135 megawatts (thermal), the NRU is a world-class facility for conducting research and producing isotopes for medical and industrial uses (see Figure 2-2).

Although neither the NRX nor NRU reactors were used to generate electricity, they were the beginning of the Canada Deuterium Uranium (CANDU) power-reactor concept. In particular, the NRU is both heavy-water moderated and cooled, and also features on-power refuelling (that is, fresh fuel can be inserted and used fuel removed without

FIGURE 2-2: The NRU research reactor at Chalk River. It first saw operation in 1957 and has been a workhorse in producing isotopes and conducting research. The reactor was still operating in 2008.

shutting down the reactor). Unlike the future CANDU reactor with a horizontal core, the NRU reactor core is vertical and fuel is loaded from the top using a transfer flask. NRU is still operating.

The NRX and NRU reactors have been veritable workhorses and are responsible for a wealth of scientific research as well as for the production of isotopes that have been used in medical and industrial facilities throughout the world (see Chapters 11 and 12). The availability of these isotopes stimulated the development of medical treatment systems including cobalt therapy units for cancer treatment. Furthermore, the NRU continues to provide engineering research and development for CANDU reactor fuel and components.

The CANDU Nuclear Power Reactor

By the early 1950s, the idea of using nuclear fission to produce electrical power was becoming an achievable goal. Based on the Chalk River developments, three companies—Atomic Energy of Canada Limited (AECL), Canadian General Electric, and Ontario Power Generation (Ontario's electrical utility)—joined forces to design and construct the Nuclear Power Demonstration (NPD) reactor. (OPG was formed from Ontario Hydro in 2000. To avoid confusion we will use OPG throughout this book even for those times when the company was called Ontario Hydro.) Ontario's electrical utility and General Electric Canada joined forces to design and construct the Nuclear Power Demonstration (NPD) reactor. Located at Rolphton on the Ottawa River just a few kilometres upstream from the Chalk River site, the 22-megawatt reactor supplied its first power to the Ontario grid in 1962.

NPD set the stage for the unique CANDU design that was to become a Canadian trademark: NPD was fuelled by natural uranium, moderated and cooled by heavy water, used thin (4 millimetre) zirconium alloy fuel channels instead of a thick (15 to 20 centimetre) steel pressure vessel typical of light-water reactors, and could be refuelled while operating. The NPD station became the training centre for nuclear staff and generated electricity until 1987, paving the way for the ambitious nuclear program subsequently implemented by Ontario Power Generation.

From that small beginning CANDU reactors can now be found producing electricity in Ontario, Quebec, New Brunswick, Argentina, India, Pakistan, Korea, Romania, and China.

The next step in developing a commercial-scale reactor was the Douglas Point reactor located on the east shore of Lake Huron, north of Kincardine. The decision to start was made in 1959, and AECL and Ontario Power Generation collaborated in the design and construction. Whereas NPD was strictly a pilot project, Douglas Point was a prototype power reactor whose purpose was to show that the small NPD reactor could be scaled up to commercial size and operated safely and economically. The power output of the Douglas Point plant was an order of magnitude larger at 206 megawatts, and placed Canada at the forefront of nuclear power alongside the United States with its Shippingport reactor and the United Kingdom with its 300-megawatt Magnox reactor.

Douglas Point went online on 15 November 1966, the first electricity was delivered in 1967, and it was declared in service in September 1968. The cost was $91 million. Owned by AECL, Douglas Point was operated by Ontario Power Generation until May 5, 1984.

Many lessons were learned from Douglas Point. The online fuelling machines proved to be troublesome and required a number of changes; there were losses of heavy water until inadequate valves and flanges were redesigned; a proprietary process for decontaminating moderator and coolant was developed; a remote-control "mouse" was developed that crawled into the bowels of the reactor and fixed a leak in 1976. The annual capacity factor—the total quantity of electricity produced in a year compared to the maximum quantity possible if the plant ran at 100 percent power for the entire year—grew from 54 percent at outset to 82 percent on retirement. In its later years, Douglas Point also supplied steam for a heavy-water plant located nearby. Douglas Point proved to be a difficult plant to operate and would have required extensive refits to keep it running economically. In 1984, new nuclear plants were being built: the four Pickering A units and four Bruce A units were already operating, and Pickering B and Bruce B were just starting to come online. The technical effort to operate Douglas Point could not be justified, and the reactor was shut down permanently. The lessons learned at Douglas Point contributed enormously to the larger stations that followed.

Ontario Power Generation was so confident with the progress made with NPD and Douglas Point that it committed to the construction of four reactors at Pickering, Ontario before Douglas Point even came

Table 2-1 Historical Development of Nuclear Power in Canada

Era	Reactor	Power	Location
1945–1970	Zero Energy Experimental Pile (ZEEP)	0	Chalk River
1947–1992	National Research Experimental (NRX)	42 MWt	Chalk River
1957–present	National Research Universal (NRU)	135 MWt	Chalk River
1962–1987	Nuclear Power Demonstration (NPD)	22 MW	Rolphton
1968–1984	Douglas Point	206 MW	Douglas Point

into service. The Pickering site is about 32 kilometres east of Toronto on the shore of Lake Ontario. The four reactors that comprise station A came online from 1971 to 1973 with a capacity of 515 megawatts each (Figure 2-3).

From 1983 to 1985 station B, consisting of four more reactors, came into operation at Pickering. These reactors have a capacity of 508 megawatts each. They were constructed adjacent to station A, so the eight reactors are in a long row. The eight units share one large vacuum building—a safety device that limits the consequences of an accident (see Chapter 8). When all eight reactors were in operation, the station generated 4,092 megawatts of electricity: this is an enormous amount of electricity, approximately double what Ontario Power Generation produces at Niagara Falls (one of the largest hydroelectric stations in Canada).

The capacity of the reactors at Pickering was insufficient to meet the continuing growth in electrical demand. Major reactor construction was commenced at the Bruce Nuclear Power Development site, located on Lake Huron about 255 kilometres northwest of Toronto. From 1977 to 1979, the Bruce A station—consisting of four reactors—came online. Each reactor supplied 750 megawatts of electricity, in addition to producing steam for heavy-water plants and the Bruce Energy Centre.

During the years 1984 to 1987, the Bruce B station came into operation with four reactors; each reactor has an 850-megawatt capacity. Unlike Pickering, the two Bruce stations are geographically separated by about 700 metres. When all eight reactors are operating, the Bruce site can deliver up to 6,400 megawatts of electricity to Ontario's grid—a prodigious amount of power. All of the reactors are of CANDU design.

With eight reactors, the Bruce site is one of the largest nuclear centres in operation anywhere and has been known as the "Nuclear

FIGURE 2-3: The Pickering Nuclear Generating Station. This station on Lake Ontario east of Toronto consists of eight reactors in two groups of four. The reactors have the rounded domes. The larger cylindrical structure is the vacuum building designed to capture and quench any radioactive steam from an accident to prevent its release to the atmosphere.

Capital of the World." It also contained heavy-water plants (all closed by 1998). The site also contains the Western Waste Management Facility where low and intermediate-level wastes from all of Ontario's nuclear generation facilities are treated and stored.

Quebec possesses much larger hydroelectric potential than Ontario; but even so, the Quebecois decided to enter the nuclear age. In 1971, the 250-megawatt Gentilly 1 reactor came into operation at Gentilly, near Trois Rivières on the shores of the St. Lawrence River. Built by AECL, the reactor was known as the CANDU-BLW, a type of CANDU reactor moderated by heavy water but cooled by boiling "light" (in other words normal, not heavy) water, rather than by pressurized heavy water. The reactor had a vertical pressure-tube core unlike all other CANDU designs, which are horizontal in orientation. Gentilly 1 was a backup design to the standard CANDU design; the reason for this was that Atomic Energy of Canada Limited was concerned that heavy-water losses at its other nuclear stations might prove to be unacceptably high—a concern because heavy water is expensive to make. Gentilly 1 was plagued by problems from the outset and was commonly referred to in Quebec to as "*le citron*" ("the lemon"). Unable to compete economically with other sources of electricity in Quebec, it was taken out of service in 1977. The reactor was used for training until it was mothballed in 1979. Heavy-water losses at Douglas Point and Pickering proved to be quite low and further development of the CANDU-BLW reactor was curtailed. However, Gentilly 2, a 685-megawatt CANDU 6 reactor (see below), was constructed at the same site and started to deliver electricity to Hydro Quebec's grid in 1983. It has operated successfully and is expected to reach its 30-year service life in 2013.

New Brunswick also produces nuclear power. A 630-megawatt reactor of the CANDU 6 type (Figure 2-4) was commissioned at Point Lepreau in 1982 and supplies about 20 percent of the province's electrical needs. It also supplies electricity to American utilities in the New England area. Whereas Ontario Power Generation's reactors were all designed as multi-unit stations, the CANDU 6 is a single, stand-alone unit designed and marketed by AECL.

Four reactors of 880 megawatts each were constructed at Darlington, about 70 kilometres east of Toronto on the shores of Lake Ontario. Begun in 1977, they were delayed for about four and a half years, primarily because Ontario's electricity demand finally slowed down. But the delay was also due to many political stoppages and labour disruptions. This led to a considerable increase in the cost of these reactors—in total about $12 billion—a result of the very high interest rates during that period. The four reactors were connected to Ontario's electrical grid between 1990 and 1993, and supply up to 3,520 megawatts of electricity in total.

The Darlington site also houses the Tritium Removal Facility (which extracts tritium from heavy water used in Ontario Power Generation's reactors). Tritium is formed by neutron irradiation of the heavy water used as moderator and coolant inside a reactor's core. More than 2 kilograms of tritium (depending on the number of reactors operating) are extracted each year from the heavy water; this is to decrease the radiation fields to which station workers are exposed. Tritium is used to illuminate exit signs and emergency markers so that they can be seen even when electrical power is interrupted. Biomedical research

FIGURE 2-4: The Point Lepreau nuclear station operated by New Brunswick Power consists of a single reactor of the CANDU 6 type

employs tritium as a tracer to track metabolism or movements of drugs and other substances in the body. Tritium is also used in fusion energy research (see Chapter 15).

These were exciting years for the Canadian nuclear industry, with 23 power reactors constructed in two decades spanning the 1970s and 1980s—more than one per year. The year 1992 was the peak of nuclear electricity production. With the completion of Darlington, there were 22 CANDU reactors operating in the country with a total capacity of 15,200 megawatts. Twenty of these were in Ontario, supplying about two-thirds of the province's electrical power in 1994. And at this time, nuclear power also supplied 19 percent of Canada's electricity. However, by 2004, nuclear's share had fallen to about 50 percent of Ontario's electricity and 15 percent for Canada. Table 2-2 summarizes the CANDU reactors in Canada.

In the latter half of the 1990s, nuclear power was to suffer badly. A dramatic decrease in the growth of electricity demand—the first time that century—cancelled plans for additional nuclear capacity. In addition, personnel cutbacks and neglect of proper maintenance by Ontario Power Generation caused significant deterioration in some of their units.

All four Pickering A reactors were temporarily shut down in 1997. Unit 4 was refurbished from 1999 to 2003, and unit 1 was back in service by 2005. That same year OPG made the decision to shut down units 2 and 3 permanently. The primary reason for this was the high costs associated with bringing units 1 and 4 back to service.

Bruce A units 1, 3, and 4 were temporarily shut down in 1997; unit 2 had been shut down previously for corrosion problems in the steam generators. In the period from 2002 to 2004, units 3 and 4 were returned to service. In 2008, refurbishment of units 1 and 2 was well advanced.

Table 2-2 Summary of CANDU Reactors in Canada

Name	No. Units	MW/unit	Start–Closure
Douglas Point	1	206	1968–84
Pickering A	4	515	1971–73
2005–ongoing			
Pickering B	4	508	1983–86
Bruce A	4	750	1977–79
Bruce B	4	850	1984–87
Darlington	4	880	1989–92
Gentilly 1	1	250	1971–77
Gentilly 2	1	685	1983
Point Lepreau	1	633	1982

In 2001, following a policy decision by the Ontario government to privatize part of the province's electricity industry, Bruce Power took over operation of the Bruce reactors under a long-term lease agreement (18 years, plus a 25-year option) with Ontario Power Generation. Bruce Power is a partner of the following: Cameco (Canada's largest uranium producer), TransCanada Corporation, BPC Generation Infrastructure Trust (a trust established by the Ontario Municipal Employees Retirement System), the Power Workers' Union, and the Society of Energy Professionals.

By 2008, with concerns about global warming and energy shortages, nuclear power came back onto the drawing board. The Ontario government announced that two new reactors would be built on the Darlington site. Another reactor was slated for New Brunswick and the possibility of power reactors for Alberta and Saskatchewan was under active consideration.

-3-

Radiation Everywhere

For some, nuclear technology and the science behind it are difficult to comprehend and therefore seem rather fearsome.

But in this respect, nuclear technology is not alone, for we live in a complex age. Our modern world is full of scientific devices that are difficult to understand and the boundaries are constantly being expanded. Today's scientists have travelled far deeper than the atom's nucleus. Now they are unravelling particles that are even smaller and more mysterious—leptons, muons, pions, and quarks. The world of science is indeed an intricate and complicated one. Who can comprehend how tiny, barely visible slivers of semiconductor chips can store millions of bits of data or compute millions of calculations per second? And think of the marvels of telecommunication, satellites and space exploration, astronomical telescopes that peer light-years into distant galaxies, organ transplants in medicine, and much, much more. It is difficult for any individual to keep abreast of all the new developments.

For those who are uncomfortable with nuclear technology and its complexity, it may be reassuring to know that radioactivity (the spontaneous emission of particles and/or electromagnetic waves from the nucleus of an atom) is an entirely natural phenomenon. Many people do not realize it, but radiation is everywhere around us, and even inside us. Radioactivity is a natural and integral part of the earth. It is as common—and as necessary—as the oxygen we breathe and the sunlight that brings life to our planet.

Before we look at nuclear technology in more detail, let us explore the role radiation plays in the natural world.

Terrestrial Radiation

Radioactive elements, primarily isotopes of uranium, thorium, and potassium, are found in all rocks and soils. They have been present since the earth was formed 4.5 billion years ago and supply the energy that drives tectonic plates to move around the globe. Where the plates grind together, mountain chains such as the Rocky Mountains, the Alps, and the Himalayas are formed.

Tectonic Plates

In the 1960s, geoscientists developed the great, unifying concept of plate tectonics to explain the distribution of mountain chains, volcanic activity, and earthquakes and to account for the accumulating evidence that continents have drifted about the globe. The entire surface of the earth consists of seven relatively thin (100 to 150 kilometres), rigid, large plates that are slowly moving about the globe. The plate margins are of considerable interest for it is here that new plate area is formed through sea-floor spreading when plates move away from each other; mountains and volcanoes form when plates move toward each other reducing plate area; and plates slip past each other without gaining or losing area. Virtually all seismic and volcanic activity takes place at plate margins. The energy that drives the movement of these plates comes primarily from the heat generated by radioactive isotopes of three elements: uranium, potassium, and thorium.

Without this ongoing process—driven by radioactivity—our planet's mountains would long ago have been eroded away by the forces of wind, rain, and snow, and the earth would be a smooth globe covered by water. Needless to say, life would have evolved completely differently. One can only speculate about how different the planet's evolutionary history would have been without this natural radioactivity.

Radioactivity is constantly decaying. From this fact, we can infer that the earth and our environment were more radioactive in the past. For example, of the uranium-235 present when the earth formed, only about 1.2 percent remains today. Although greatly diminished, radioactivity continues to shape the earth, and is an important tool that geoscientists use to date the age of different rocks and to learn about the history of the earth.

The radioactivity from rocks enters soils, building materials, and virtually everything on earth. Plants take up radioactivity from the soil through their roots and, in turn, humans ingest radioactivity when we eat. Soils also emit radon-222, a radioactive gas which arises from the uranium-238 decay chain (see Appendix A), and radon gas becomes a component of the air we breathe. All rocks, all soils, all plants, and all living things, including humans, contain natural radioactivity. In fact, it is impossible to find or construct a place on earth that is completely free of radiation.

Potassium-40, thorium-232, and uranium-238, known as "primordial radionuclides" because they have been present since the earth was formed, are the most significant sources of radiation in the environment. As discussed in Appendix A, each of uranium-235, uranium-238, and thorium-232 has a long half-life (about a billion years or more) and decays through a long series of radioactive "daughters" (decay products) until a stable isotope of lead is reached.

Neither uranium nor thorium themselves are particularly hazardous for human health because neither participates to any significant degree in plant or animal metabolism. Some of their daughters, however, are important, particularly radium-226 and radium-228, because they bear a chemical similarity to calcium.

Table 3-1 shows six of the 18 radionuclides that occur on earth and do not arise from the uranium or thorium decay chains. These have very long half-lives and many are exotic elements with high mass numbers. Most of the elements occur as mixtures of isotopes. For example, potassium has nine isotopes, ranging from potassium-37 to potassium-45. Six of these isotopes decay rapidly and are not found in nature. Two of them are stable (potassium-39 and potassium-41) and together with the small percentage of radioactive potassium-40, make up naturally occurring potassium. Rubidium-87, for example, has almost the same abundance as potassium-40 but is of minor interest: the beta radiation it emits as it decays to strontium-87 is very weak. In contrast, potassium-40 emits a very energetic gamma ray.

Table 3-2 shows typical naturally occurring uranium concentrations. The oceans are characterized by low concentrations of uranium (and the other main radioactive nuclides). When our distant ancestors emerged from the sea to crawl and eventually to walk on land,

Table 3-1 Some Primordial Terrestrial Radionuclides

Isotope	Half-life
Potassium-40	1.27×10^{9} years
Vanadium-50	4×10^{16} years
Rubidium-87	4.8×10^{10} years
Indium-115	5×10^{14} years
Tellurium-123	1.2×10^{13} years
Lanthanum-138	1.1×10^{11} years

Table 3-2 Typical Naturally Occurring Uranium Concentrations (Parts per million Uranium)

High-grade ore body (10% uranium)	100,000
Low grade ore body (0.1% uranium)	1,000
Granite	4
Sedimentary rock	2
Average in continental crust	1.4
Seawater	0.003

they entered an environment with hundreds of times more radiation.

The amount of uranium and thorium in different rock types varies considerably from place to place depending on the local geology. In some areas the radioactive elements have been concentrated by natural processes, sometimes to the point where it is worthwhile to mine them. Uranium ore bodies are found in Canada (see Chapter 13) as well as in Australia, Russia, the United States, and other countries. The average concentration of uranium in these ore bodies, which are found from near the surface to several kilometres deep, ranges from about 0.1 percent to over 15 percent. These ore bodies provide considerable information on how radioactive materials behave in our environment and the impact that they have on human health.

When uranium and thorium are contained in the minerals zircon and monazite, they are particularly resistant to chemical weathering. Prolonged wave action on beaches can winnow out less dense quartz and other minerals, leaving the more radioactive elements. Significant concentrations of these radioactive minerals are known in Brazil and southwest India. The latter contains the highest thorium concentrations (8 to 10 percent) in the world.

Other areas with high terrestrial radioactivity include western Australia, southwest England, and southeastern United States. High radium levels are found in waters in Illinois and Iowa. The most remarkable instance of high natural radiation levels is in Ramsar, Iran, where there are up to 50 hot springs delivering large amounts of radium into the environment. The inhabitants of these places are exposed to considerably more radiation than people living elsewhere.

In summary, natural radiation is quite variable depending largely on the local geology. People living in areas of limestone receive lower doses of radiation than those living in areas with more granite. In extreme cases, such as in one area of Brazil where people live on soils containing abnormally high concentrations of thorium, the external radiation dose can be up to 120 times greater than found in average areas. As discussed in Chapter 4, people living in such areas show no adverse health effects due to high radiation.

Radon

Radon merits separate discussion because it is the largest component (approximately half) of the natural radiation to which we are exposed. Radon is a colourless and odourless gas formed from the radioactive decay of radium, which in turn is one of the daughter products of the uranium decay chain. In the uranium-238 series, radium-226 decays to radon-222, which has a half-life of 3.8 days (see Table A-2 in the Appendix). Radon is also produced in the thorium-232 series, where radium-228 decays, via some intermediate daughters, to radon-220. Since this has a half-life of only 53 seconds, it generally decays before escaping from the soil and is not as significant a concern as radon-222.

Radon is constantly being produced and released from the ground and is always present in air. Outdoors it is greatly diluted by fresh air and is generally found in low concentrations that are not a health concern. It can, however, seep into and collect in buildings and become more concentrated. Seepage into a house takes place through cracks, joints, dirt floors, and around drainage pipes and sump pumps. Indoor levels of radon can vary greatly depending on the location of a building and how it is sealed and ventilated.

It is not the radon itself that is a health concern, but the radioactive daughter products into which it decays. Radon, being a gas, is simply the vehicle by which members of the uranium decay chain can escape from the soil and enter our lungs. Outside the body, radon is not a concern since alpha particles cannot penetrate the skin. Furthermore, being chemically inert (it is one of the noble gases), radon can be inhaled and exhaled harmlessly. However, radon decays into daughter products that are solid and, having an electrical charge, can attach themselves to dust particles. In this form, the radon daughters can lodge in the bronchial and lung tissues where they emit alpha particles that can damage surrounding cells and, if concentrations are high enough, could cause cancer.

Radon is the subject of considerable research and debate. The US Centers for Disease Control and Prevention believes radon to be the second greatest cause of lung cancer, next to smoking, and estimates it may cause between 5 and 10 percent of all lung cancers. Because of this concern, the Surgeon General of the United States urged that most homes be tested for radon levels. However, a study by Cohen (1995) that included 90 percent of the population of the United States showed that those counties with higher radon concentrations in fact had lower lung-cancer mortalities.

Health Canada does not recommend testing of homes although it has recently revised its voluntary guideline from 800 to 200 becquerels/cubic metre (see Appendix A for an explanation of these units) for the maximum acceptable level of radon. For perspective, if a house with radon concentrations of 200 becquerels/cubic metre were occupied 70 percent of the time, the occupant would receive a dose a few times that of natural background. Based on measurements across Canada, it is estimated that less than one home in a thousand requires corrective action.

The radon level in a home can be measured by activated-charcoal detectors. These consist of containers that allow air to enter, and then radon and its daughters are absorbed by the charcoal. These detectors are generally exposed for a day or two and then sent to a laboratory for analysis. They are easy to use and inexpensive. Due to the brief sampling period, they are best used for screening to determine if more detailed studies are needed.

Alpha track detectors, which use a sheet of polycarbonate plastic in a container, are also easy to use. The radioactive emissions of alpha particles cause small damage marks in the plastic which are subsequently measured with suitable instruments. The container is generally left in place for at least three months.

Radon levels in homes can be lowered relatively simply and inexpensively by improving circulation and ventilation and by sealing cracks in the foundation and around pipes and sump pumps.

Cosmic Radiation

In 1911, the British scientist C.T. Wilson invented a device now called the Wilson Cloud Chamber, which allows radiation to be seen. A cloud chamber experiment is so simple that it can be done in a high-school laboratory or even at home. A clear glass or plastic container forms the chamber, and ethyl alcohol, super cooled by dry ice, is used to form a "cloud." As radiation passes through the chamber, it knocks electrons off—a process otherwise known as "ionizing"—the atoms in the air. The alcohol vapour then condenses on the charged particles, forming tracks similar to the vapour trails of jet planes (See Figure 3-1). These are easily visible by shining a light through the chamber in a darkened room. The tracks disappear relatively quickly. Most of the tracks will be about one centimetre long and are quite distinct; these are caused by alpha particles. Beta particles cause longer, thinner tracks. Occasionally, faint, twisting, circling tracks are observed; these are due to gamma rays.

The cloud chamber confirmed that the earth is being bombarded by intense radiation originating from space. Although it took decades to understand the processes involved, it was soon realized that nuclear reactions drive the universe, providing the fundamental power to make stars shine and give light and warmth to planets like the earth. All stars, including our sun, are giant nuclear reactors powered by a process called nuclear fusion (described in Chapter 15). Not only does the sun create the light and heat on which our world depends, but the giant inferno inside the sun is constantly ejecting a stream of energy and particles, called the solar wind, into space. The amount of this cosmic radiation increases considerably during solar flares, when sun spots shoot solar material into outer space like volcanoes.

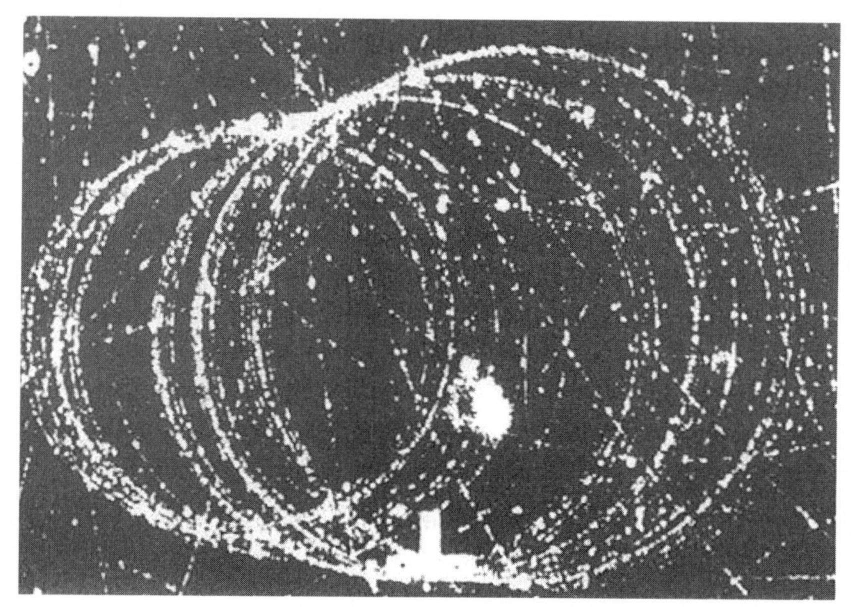

FIGURE 3-1: Cloud chamber showing vapour trails caused by cosmic radiation. Some trails are circular because a magnetic field was applied for this photograph to help in identifying the masses of the particles.

The solar wind travels through space and its particles react with the earth's atmosphere creating cosmic radiation that constantly rains down on us. The solar wind consists largely of protons together with alpha particles, a variety of heavier particles, electrons, photons, and neutrinos. Although most of these particles come from our sun, some originate in far-off suns and galaxies. The solar wind interacts with the upper atmosphere to ionize some gases and also to create new radionuclides. This radiation extends from the upper atmosphere all the way to the earth's surface. So all living beings are constantly bombarded by millions of particles of cosmic nuclear radiation, whether they live on earth (or, for that matter, anywhere else in the universe). Table 3-3 lists some of the radionuclides that have a cosmic origin.

The amount of cosmic radiation is the lowest at sea level because of the shielding provided by the atmosphere; it increases with altitude. Residents of Banff, for example, receive 0.2 millisieverts (mSv) per year (see Appendix A for an explanation of millisieverts), more radiation than the inhabitants of Halifax. Travelling in an airplane yields more radiation than remaining on the ground. For example, a cross-country

Table 3-3 Radionuclides of Cosmic Origin

Isotope	Half-life	Production Rate*
Hydrogen-3 (tritium)	12.3 years	0.25
Beryllium-7	53.6 days	8.1×10^{-3}
Beryllium-10	2.5×10^{6} years	3.6×10^{-2}
Carbon-14	5730 years	2.2
Sodium-22	2.6 years	5.6×10^{-5}
Sodium-24	15 hours	**
Silicon-32	650 years	1.6×10^{-4}
Phosphorous-32	14.3 days	8.1×10^{-4}
Phosphorous-33	24.4 days	6.8×10^{-4}
Sulphur-35	88 days	1.4×10^{-3}
Sulphur-38	2.87 hours	**
Chlorine-36	3.1×10^{5} years	1.1×10^{-3}
Chlorine-38	37 minutes	**
Chlorine-39	55 minutes	1.6×10^{-3}

* in atoms per square centimetre per second
** half life too small to measure production rate

flight yields an additional 0.02 to 0.03 millisieverts of cosmic radiation than if one had stayed at ground level. Commercial aircraft crew, who typically log about 700 hours of flight time in a year, receive about 2 to 5 millisieverts per year of radiation from flying (the average Canadian is exposed to about 2.6 millisieverts in one year).

Astronauts who travel in space on lunar visits or to orbit around the earth receive even more radiation, particularly if they stay "up" for an extended period of time. On the Apollo XIV mission to the moon in January 1971, it was estimated that the astronauts received a radiation dose of approximately 5 millisieverts in 12 days (about two annual background doses). This was a real concern when the NASA space program first started, and considerable research has been done to study the effect of cosmic radiation not only on astronauts but also on electronics and other delicate control systems used in space. The only effects on astronaut health appear to be an increase in cataracts. No increased cancers have been detected, but the astronaut population is still too small and too young to obtain robust statistical data.

A particular concern continues to be the huge increase in cosmic radiation that occurs when the sun has solar flares. Space missions are scheduled to avoid solar flares, whenever possible. If humans were to travel to distant solar systems, as they do in science-fiction, they would be exposed to elevated levels of cosmic radiation unless some form of shielding is provided.

Radiation inside the Body

Since there is so much radiation in the environment, it is not surprising that there is also substantial radiation in our bodies. Potassium-40 is the most important radioactive component of food and human tissue. About 120 out of every million atoms of potassium (0.012 percent) are radioactive; the rest are stable. Potassium is a component of soil and from there is taken up by plant roots. It comes into our bodies directly when we eat fruits and vegetables and indirectly via the meat of animals who eat root crops. The radioactive potassium is then deposited in parts of our body such as bones. Potassium helps maintain fluid pressure and balance within cells. It is also important in normal muscle and nerve response and in maintaining heart rhythm. Even a temporary potassium deficiency can result in serious upsets of body functions.

Scientists have established that the potassium-40 in our bodies undergoes about 60 decays per second per kilogram of body weight. This means there are approximately 4,500 radioactive decays per second, each emitting a gamma ray, that take place inside a person weighing 75 kilograms (165 pounds). Combined with other natural radioisotopes inside and outside our bodies, this person is struck by nuclear radiation about 54 million times in a single hour. Every single day of our existence over a billion radioactive particles pass through our body. Through the long evolution of humans, our bodies have adapted to this radioactivity.

Carbon-14 is another important biological element, since carbon forms about 23 percent of our bodies by weight. Carbon is a major component of the food we eat, forming about 45 percent of carbohydrates, about 55 percent of fat, and about 50 percent of proteins. Carbon-14 is responsible for almost as much radioactivity within our bodies as potassium-40 (about 3,900 decays per second for a person weighing 75 kilograms).

It is worth remembering that radioactivity is not only a natural part of our environment, but also of our bodies.

Human-Made Radiation

Some of the radiation in the environment arises from human activities. The largest human-made source of radiation is from medical applications. Other very small contributors are nuclear laboratories, industrial and consumer sources such as smoke detectors, and atmospheric fallout from nuclear weapons testing. In addition, nuclear power reactors contribute less than 0.04 percent of the radiation in the environment (CNSC, 1995).

About 90 percent of medical radiation doses come from X-rays, with each person in Canada receiving, on average, 1.2 X-rays per year. Of these, 14 percent are dental X-rays. Technically, the radiation from X-rays does not emanate from the nucleus, but is caused by transitions in the electrons around the nucleus. Thus, X-rays are an electromagnetic, not a nuclear, phenomenon. The effects of X-rays, however, are the same as those caused by nuclear radiation. Other medical radiation doses come from radioactive isotopes used in various diagnostic tracer tests involving small doses.

Coal-fired power plants release radiation in their emissions due to the radioactive elements naturally contained in coal. Radiation can arise from the release of radon from disturbing the earth during construction and road-building projects, and from the use of phosphate fertilizers, which contain relatively high concentrations of natural radioactive elements.

Consumer products that emit very small amounts of radiation include colour televisions (those with cathode-ray tubes) and smoke detectors.

From 1945 to 1963, the Soviet Union, the United States, the United Kingdom, and France developed nuclear weapons, and tested them by detonating them above ground. These tests sent radioactive materials high into the atmosphere where they were dispersed by winds and carried around the globe. The radionuclides of principal biological concern were tritium (12.3-year half-life), strontium-90 (28 years), cesium-137 (30 years), iodine-131 (8 days), manganese-54 (314 days), iron-55 (2.7 years), and cobalt-60 (5.3 years). By the late 1950s, the increase in atmospheric radiation was measurable and was causing alarm.

On August 5, 1963, the Soviet Union, the United States, and the United Kingdom signed the Treaty Banning Weapons Testing in the Atmosphere, in Outer Space, and Under Water, known more familiarly as the Limited Nuclear Test Ban Treaty. Atmospheric bomb testing was banned, and subsequent tests were conducted underground. With the collapse of the Soviet Union and the end of the Cold War has come a significant and very welcome movement away from nuclear weapons. This has included a decrease in the testing of nuclear weapons and also in the size of nuclear arsenals. The Nuclear Non-Proliferation Treaty, signed in 1968, was extended for an indefinite period in 1995. In addition, the Comprehensive Nuclear Test Ban Treaty (including underground tests) was adopted by the United Nations in 1996 and has now been ratified by 144 nations. Tests in the Pacific and elsewhere have ceased. Latin America and Africa are nuclear weapons–free continents, and the United States and the Russian Federation downsized their nuclear arsenals. An unfortunate development that ran counter to this trend was the underground bomb tests conducted by India and Pakistan in 1998.

The average annual dose from atmospheric weapons testing conducted between 1945 and 1963 has decreased from a maximum of about 0.15 millisieverts per year in 1963 to less than 0.005 millisieverts per year today. The latter amount represents a negligible contribution (less than one quarter of one percent) to natural radiation.

The notorious explosion of the Chernobyl nuclear reactor in the Ukraine on 26 April 1986 released large amounts of radioactivity into the atmosphere (see Chapter 8). Air currents carried the cloud in two directions, reaching Canada's east coast on May 6 and the west coast on May 7. The highest air concentrations in Canada occurred in May and by the end of June the levels had returned to normal. Health Canada estimated that the Chernobyl accident led to the typical Canadian receiving a radiation dose of 0.00028 millisieverts—a negligible amount compared to typical variations in natural background.

The reality is that we live in a sea of natural radiation. The amount of radioactivity any specific individual is exposed to varies with location, altitude, diet, and lifestyle. On average, each Canadian receives approximately 2.6 millisieverts of radiation per year, including medical treatments. This amount of radiation is a useful yardstick and will be referred to as the "annual background dose."

The sources of background radiation are divided into those of cosmic, terrestrial, and internal origin. Radon gas, although of terrestrial origin, is usually discussed separately because it makes such a significant contribution. The contributions to total background radiation are summarized in Figure 3-2, which shows that about 76 percent (2.0 millisieverts) is from natural radioactivity and about 23 percent (0.62 millisieverts) is from medical, dental, and other sources. The category "Other" includes consumer products, fallout from nuclear weapons testing, nuclear power reactors, and uranium mining (CNSC, 1995).

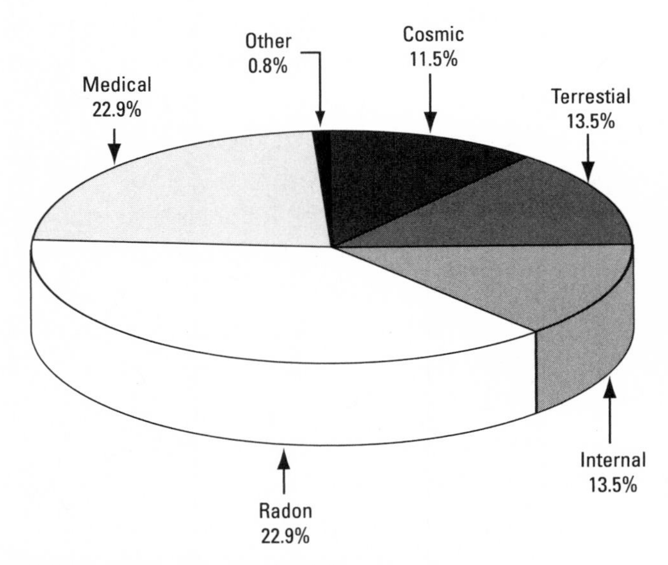

FIGURE 3-2: Sources of average annual radiation exposure.

-4-

Biological Effects of Radiation

We have always been immersed in radiation. Our bodies are radioactive, as are the rocks, plants, and air around us; it is a natural and unavoidable part of the world.

As a typical adult is "struck" by about one billion individual radiations each day, it is clear that the cells that compose our bodies can cope with and, indeed, thrive in this radiation field. This is supported by the fact that there is considerable variation in natural radioactivity with some localities having significantly higher radiation than the average. Yet extensive studies have shown that the people in these areas are just as healthy as those who live elsewhere.

Nevertheless, larger exposures to radiation can be damaging to health, as for example, noted in the early 1900s when radium dial painters inadvertently swallowed appreciable amounts of radium when they licked their brushes to make a sharper point. In extreme cases, a very large amount of radiation can cause death, although such a large accidental (acute) exposure of radiation has never happened in Canada. It is for this reason that questions in radiation biology do not focus on large doses. Instead, they are directed at revealing what effects small amounts of radiation have on humans. This has important ramifications: for example, answers to these questions would assist in setting allowable levels of radiation for nuclear workers. Unfortunately, the questions are not easily answered, and in spite of considerable efforts there is some debate on this subject.

This chapter focuses on what are called "low levels" of radiation, that is, radiation doses from 0 to 100 millisieverts. It should be noted that even though these doses are termed "low level," the upper end of this range greatly exceeds virtually any radiation levels to which Canadians would ever be normally exposed. The natural background radiation delivers on average an annual dose of around 2.6 millisieverts. This background dose is at the low end of the range referred to as low-level radiation.

Numerous statistical comparisons have been made of groups of people who have received low levels of radiation. No clear evidence has yet been found that such radiation has an adverse effect on human health. Nevertheless, scientists continue to search for low-level radiation effects not only because of the possible implications for health but also because of information they reveal about fundamental aspects of cell biology, for example, DNA repair.

There are two different ways of approaching this topic. One is to study radiation effects on the microscopic level to see how ionizing radiation interacts with the cell and different molecules within it. The other approach is to look at the problem from a larger perspective in terms of the effects of radiation on populations exposed to different amounts of radiation. Let us look at the former approach first.

The Microscopic View

Measuring Radiation: Dose Units

To discuss the biological effects of radiation meaningfully, it is helpful to understand the units in which radiation dose is measured. These were mentioned in Chapter 3 (and are also discussed in Appendix A), and will be described in more detail here.

Radiation deposits energy in the material through which it passes. The energy absorbed by the material is measured using a unit called the "gray" (1 gray equals 1 joule of energy absorbed per kilogram of matter). Thus, the radiation dose to humans could be measured in grays.

If an alpha, a beta, and a gamma each deposit one gray of energy in human tissue, will they have the same biological effect—that is, will they cause the same amount of damage? The answer is no. The alpha particle (helium nucleus), due to its size and charge, travels the least distance; so it deposits its energy over the smallest area. Because this energy is very condensed, spread over a millimetre or less, it causes more damage than the same amount of energy deposited by a gamma, which may be spread over a metre. To allow the biological effects from different radiation to be compared in a consistent manner, a new unit was developed called the sievert (Sv). A sievert is defined as a gray multiplied by a quality factor, Q, which measures the relative biological effectiveness of the different types of radiation.

The values of Q are:

Q = 1 for gamma rays and beta particles
Q = 3 to 10 for neutrons (depending on neutron energy)
Q = 10 to 20 for alpha particles

In other words, the sievert (Sv) is a measure of the energy deposited by radiation in the human body, adjusted for different types of radiation causing different damage. As doses encountered in real life are much smaller than a sievert, the unit millisievert (i.e. one-thousandth of a sievert) is generally used. Typical doses that you, I, and even workers in nuclear power plants are exposed to range from a fraction of one millisievert to a few millisieverts.

There is another complication. In many situations the radiation to a person is not uniform but instead is concentrated in a particular organ, such as the lungs or the hands. Different organs respond differently to the same amount of radiation. A given dose to the lungs would be quite different from the same dose to the thyroid—lungs are 40 times more sensitive to radiation than the thyroid. To allow meaningful comparisons, it is useful to convert an organ-specific dose to an equivalent dose that would be delivered to the whole body to achieve the same effect. In this book we refer only to the whole-body effective dose, even if the dose is to a specific organ.

There is no doubt that radiation can cause damage to living organisms. A schematic illustration of this process is shown in Figure 4-1. As radiation passes through human tissue it knocks electrons out of their electronic shells, leaving a trail of ion "pairs." That is, the molecule with the removed electron is positively charged, and the molecule to which the removed electron attaches itself becomes negatively charged. These ionized molecules, i.e., those with electrical charges, are chemically very reactive (in other words, there is a strong impetus to form a new molecule). The resulting chemical changes can upset the natural chemical reactions that take place in living cells and can cause biological effects.

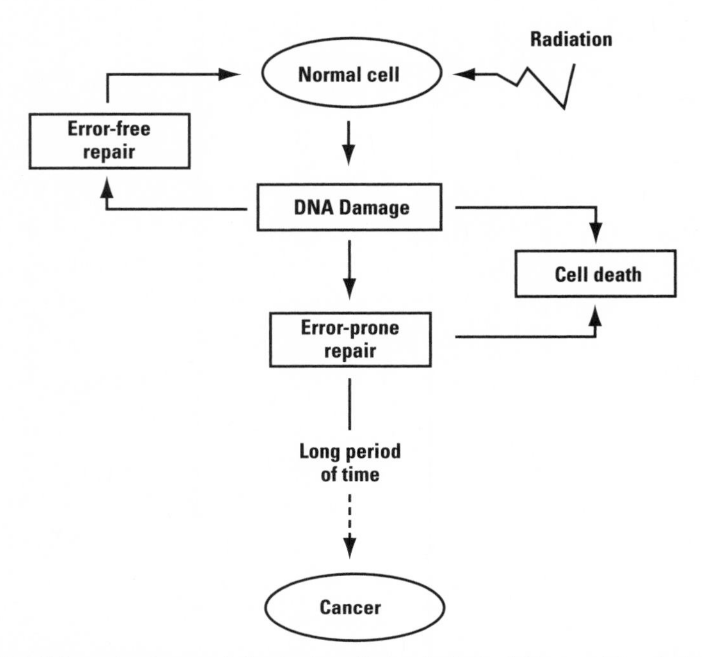

FIGURE 4-1: Schematic of cancer induction by radiation. Radiation initiates a defect in the DNA of a cell. Usually the damage can be repaired by the cell's defence mechanisms but sometimes the damage is such that the cell cannot survive or the cell dies as part of a natural mechanism that cleans up defective cells. In rare cases the repair contains errors which transform the cell into a cancer cell.

Gamma rays can penetrate deep into the body and cause damage even at a distance from the source of radiation. Beta particles can only penetrate a short way into tissues; but even so, they can cause skin burns and damage if ingested. Because alpha particles cannot penetrate the skin, they are a health concern only if they are carried inside the body by breathing or eating.

The effect of radiation on the human body depends on the duration of the exposure, how much energy is deposited in the body (i.e., the strength and type of radiation), and the type and number of cells that are affected. The body has defences against radiation; there are important repair mechanisms within cells.

Through studies of 96,000 survivors of the 1945 atomic bombings of Japan, scientists have learned a great deal about the biological damage caused by large radiation doses received over a short time (known as acute exposure). No immediate adverse health effects have ever been observed for acute whole-body radiation doses less than 200 millisieverts (ICRP, 1990; UNSCEAR, 1993). In large doses, ionizing radiation can cause death or damage to cells and cell components and, when sufficiently large, they can cause the organism to die. For example, a radiation dose over 1,000 millisieverts would usually cause radiation sickness; a dose of 4,000 millisieverts can cause death. But such acute exposures are extremely rare. The two incidents of modern times are the doses suffered in the atomic bombs dropped on Hiroshima and Nagasaki, as well as by some firefighters and reactor operators present at the Chernobyl accident. No acute exposures of this magnitude have ever occurred in Canada by accident (although controlled large doses are used in cancer therapy).

Chronic exposure consists of low levels of radiation over long periods of time. For such chronic doses, there are no immediately observable effects. Any effects, if they do occur, appear many years later. For this reason they can be difficult to link to the original cause. Because of this uncertainty, considerable controversy has arisen over low doses of radiation.

Although radiation can cause minor damage to other molecules, it can have a significant impact only on a special molecule called DNA, which contains information that controls how cell behaves (i.e., genetic material). Ionizing radiation can cause damage to the DNA molecule that may affect its future behaviour. It is unrepaired or incorrectly repaired cell damage that is critical, rather than cell death. A dead cell is replaced, but a damaged cell may replicate itself. For example, if the cell multiplies out of control, it may become a cancer tumour. Organs with rapidly dividing cells (bone marrow, gonads, intestines, growing children, fetuses in utero) are most susceptible to radiation damage. For this reason, pregnant women are not allowed to work within the radiation-containing areas of nuclear facilities.

Somatic effects are those effects that appear in the exposed person. Genetic effects are those that appear in children after the parent has been exposed.

The main adverse effect of radiation on humans is cancer, which does not generally become evident until many years or decades after the exposure. Extensive studies with survivors of the atomic bomb

Risk

No activity is perfectly safe; everything in life carries a risk, whether it is driving a car, crossing a street, smoking a cigarette, climbing a mountain, or enjoying a swim. When regulatory agencies such as Environment Canada, Health Canada, the Canadian Nuclear Safety Commission, and the US Environmental Protection Agency set limits on, for example, the maximum amount of certain heavy metals allowed in drinking water, they do so by calculating the risk that ingestion of that chemical would cause. The maximum concentration is set so that the resultant risk of death or serious health damage is very small (it is impossible to make it zero). The risk number is determined by comparison to other risks that humans accept, such as the risk of an accident on an airplane trip, and so on. A standard that has evolved in recent years is that such risks should be kept smaller than one chance in 100,000 over a lifetime.

To get a perspective on this, imagine 100,000 thimbles on a (large) table; only one thimble has a pea under it. Given only one choice, the chance of your selecting the right one is one in 100,000.

Some typical risks in everyday life are as follows. The risk of death from a driving accident is one in 5,000 per year or one in 100 over a 50-year period. The risk of death per year from an accident at home is one in 11,000. The risk of death per year from an accident at work is one in 24,000. The probability of hitting the jackpot in Lotto 649 is about 1 in 14 million.

detonations in Japan show that no other diseases are caused by radiation. In fact, at low doses, no cancer has ever been positively correlated as being caused by radiation. Many people do get cancer, however, and it is thought that some small fraction of those may be a result of radiation.

If the damaged cell is a sperm or egg cell that subsequently develops into a fetus, hereditary effects may occur. Although this has been observed in animals, radiation-caused genetic defects have never been observed in humans, despite extensive studies. For example, no significant increase in genetic disorders has been detected in the children of the Japanese bomb survivors at Hiroshima and Nagasaki.

Laboratory Experiments

One might think that exposing bacteria, laboratory animals, and human cell colonies to low-level radiation in the laboratory might provide answers. Some experimental results have been as follows:

- Bacterial colonies exposed to three times the natural level of radiation thrive much better than those exposed to less than the natural background level.
- Various strains of mice exposed to very low doses of gamma radiation live longer and have fewer cancers than non-irradiated mice. The survival time of 50 percent of the exposed mice was 673 days, compared with 549 days for the control group. Furthermore, exposed mice had fewer cancers than their non-exposed counterpart.
- Tests with cell colonies show that at an average of one radiation hit per cell (approximately the number of radiation hits per year in human cells at natural radiation levels) the risk of transformation of a cell to a cancer cell was reduced three to four times below the rate of transformation when there was no radiation. The data also show that even higher doses—up to 100 times greater, but delivered at a slow rate—produced the same reduction in cancer transformation risk.

These results are unexpected.

Search for a Risk-Dose Relationship

In view of these laboratory findings, there is some controversy among scientists about the effects of low exposures of radiation. Virtually all debate about the health effects of radiation centres about what is called the "linear no-threshold" dose theory. It may seem a little complicated at first, but provides a fascinating insight into statistics and science.

Ideally, medical scientists would like to know the exact relationship between dose and risk: if a person receives a certain radiation dose (in millisieverts), what is the risk of developing cancer (usually expressed in probability over a lifetime)?

Such a relationship would be valuable in determining regulatory standards and generally assisting in radiation protection. Put otherwise, standards would be set for nuclear facilities and practices so that the doses to their workers and to the public would not exceed an acceptably small risk (usually about 10^{-5} to 10^{-6}, or somewhere between a single chance in 100,000 and one chance in 1 million). In spite of decades of research, it has proven very difficult to determine such a dose-risk relationship.

Figure 4-2 shows Cancer Risk plotted against Radiation Dose. As expected, the higher the dose, the greater the risk of developing a form of cancer. The data to plot this graph have all been obtained from high doses, primarily from the nuclear bomb detonations at Nagasaki and Hiroshima, the reactor accident at Chernobyl, early X-ray workers, early radium dial painters, bomb testing at Bikini Atoll, and also from animal tests.

The curve for the low-dose section in which we are interested (0 to about 100 millisieverts) is not known. It is exceedingly difficult to obtain reliable dose-risk data for this part of the graph because as we have seen any biological effects of low radiation dose, if they occur, do so many years after the dose has been received. Cancer is primarily a disease of old age, and may occur 10 to 40 years after the dose has been received. Furthermore, if a cancer does appear, it is impossible to separate radiation-caused cancers from those generated by other causes. More than 1,500 different agents other than radiation are known to cause cancer—arsenic, tobacco smoke, asbestos, ultraviolet light, paraffin oil, fungal toxins in food, viruses, and even heat.

Because of the difficulty in accurately correlating dose and small effects, it is easy to misinterpret data. Suppose, for example, that 20,000 cancer deaths are predicted in a population of 100,000 people. What is the significance if 20,300 deaths were actually to occur? Does this mean that radiation, or some other cause, led to 300 excess deaths? Or is this simply part of the natural statistical variation? This uncertainty would have been removed if the predicted cancer deaths had been quoted together with the expected variation, as 20,000 plus or

Table 4-1 Cancer Deaths and their Causes (Colditz et al., 1996)

Cause	Cancer Deaths (%)
Diet	30
Tobacco	30
Sedentary lifestyle	5
Occupational factors	5
Family history of cancer	5
Viruses/Biological agents	5
Perinatal factors/growth	5
Reproductive factors	3
Alcohol	3
Socioeconomic status	3
Environmental pollution	2
Ionizing/UV radiation	2
Prescription drugs/medical procedures	1
Salt/food additives/contaminants	1

minus 1,000. Because of this type of statistical variation, it is possible to "prove" almost anything. Data of this nature need to be carefully reviewed by experts in statistical analysis and epidemiology.

In western countries about 20 percent of all deaths are due to cancer, with the average cancer death occurring at about 65 years of age. Table 4-1 shows the various causes of cancer deaths. It is seen that 3 percent of cancer deaths are attributed to the natural environment, which includes natural radiation. Diet (35 percent) and smoking (30 percent) are the main causes of cancer deaths in the United States and in Canada.

For the preceding reasons, the shape of the dose-risk curve for doses below 100 millisieverts is not known. Nevertheless, various hypotheses have been proposed. Four different forms of the dose–risk relationship are discussed below and illustrated in Figure 4-2.

a. The most widely used theory assumes a straight line all the way from high doses to zero dose, as shown by curve "a" in Figure 4-2. This is the so-called "Linear No-Threshold" model. The line is based entirely on data obtained at doses that are far outside our region of interest (in other words, not any sort of amount a Canadian will ever encounter). The slope of the line yields the risk per dose for the total population, which is 0.005 percent cancer risk per millisievert of dose (ICRP, 1990). The linear theory has one very

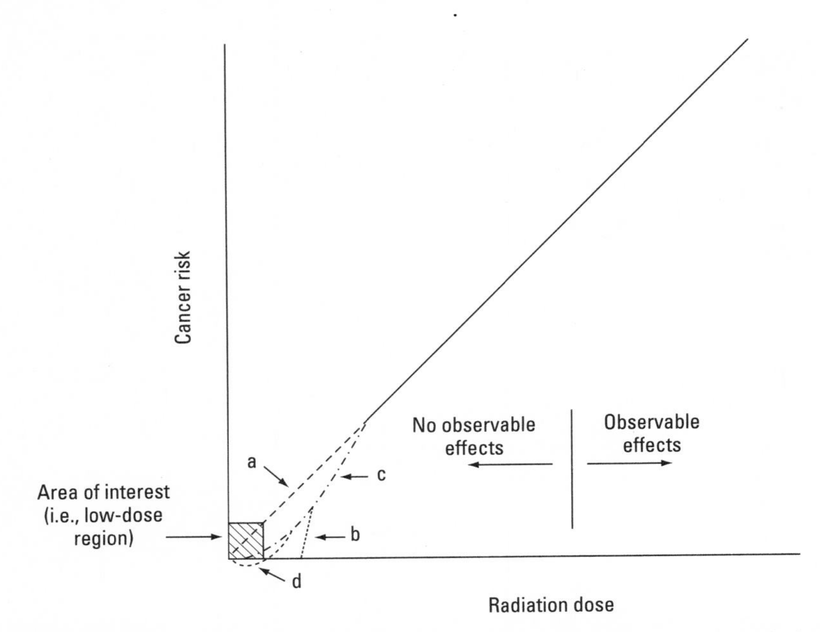

FIGURE 4-2: Curves illustrating various radiation-dose versus cancer-risk models used for low radiation exposures: (a) linear, (b) threshold, (c) S-shaped, and (d) adaptive response.

important implication: all radiation is harmful, no matter how small the amount. (Note that this has yet to be proven, and in fact there is research underway now exploring the very real possibility that higher-than-usual doses of radiation can improve health, perhaps because such a dose stimulates cell repair. See below.).

It is important to stress that the majority of scientists and doctors who have studied the subject believe that low levels of radiation have an extremely low risk, if any, of a harmful effect. In fact, virtually all expert scientific groups recognize that the linear no-threshold theory overestimates risk. For example, the Committee on the Biological Effects of Ionizing Radiations of the US National Research Council, the United Nations Scientific Committee on the Effects of Atomic Radiation (1988), and the US National Committee on Radiological Protection and Measurements (1980) have included a dose and dose-rate effectiveness factor of about 2, which halves the risk per unit dose at low doses or low dose rates (or both) compared to the risk given by a linear extrapolation from the high dose region. The US Committee on Biological Effects of Ionizing Radiation (1990)

Epidemiology

Epidemiology is the science of epidemics, or the study of diseases that attack many people. It is particularly useful when there is no readily observable relationship between the cause and symptom, so that large numbers of people must be studied, generally over a long period of time, using computer databases and statistical analyses.

also acknowledges that the linear theory overestimates risk, but continues to use linear extrapolation.

Although the linear theory does not match what has actually been observed, it is used by regulatory agencies to set radiation dose regulations because it is so conservative. That is, regulations based on the linear theory will have an extra factor of safety. The linear theory also has the advantage of being simple. For example, even though many scientists believe that the actual relationship includes a threshold (curve b), determining the specific value for this threshold is very difficult.

b. The most likely possibility is that there is some threshold value of dose below which there is no risk, as shown by curve "b" in Figure 4-2. Only doses that exceed the threshold value cause cancer risk. This curve matches evidence that our bodies can repair and cope with small amounts of damage. It is also consistent with how our bodies deal with many other substances. For example, molybdenum and iron are vital for health in trace amounts, yet are poisons if taken in large amounts. Another example is a common headache tablet such as aspirin, which if taken in large quantities can be lethal, but if taken by ones and twos eases pain and headaches. Furthermore, an exhaustive study of workers who painted radium dials was conducted over 40 years and showed there was a dose threshold (Rowland, undated).

c. Another possibility is that the risk/dose curve lies somewhere between the linear and threshold curves. This is a compromise between the linear and threshold theories. The curve could take many forms of which the S-shaped curve (curve "c" in Figure 4-2) is used relatively often. This curve indicates that the risk at low doses is quite small but is not zero.

d. A fourth alternative is that low radiation doses can actually have a beneficial effect. This effect is called "adaptive response" or "hormesis" and is similar to immunization and vaccination, where small doses provide protection against larger doses. A substantial body of evidence is accumulating that cells exposed to low-level radiation suffer less damage later when exposed to larger levels of radiation. For example, it has been observed that Japanese bomb victims who received only small radiation doses are developing fewer cancers and living longer than the control group who received no bomb dose. Further evidence for the validity of the hormesis effect comes from Taiwan. The cancer mortality rate for a large group of people who had lived up to 20 years in a group of 180 apartments in which the reinforcing bars in the concrete had been contaminated by cobalt-60 was found to be less than 3 percent of that of the general adult population (Chen, et al., 2004). The homesis effect has been traced to stimulated production of repair enzymes by low-level radiation. Adaptive response is illustrated by curve "d" in Figure 4-2.

Let us see what would happen if the linear theory were applied to headache tablets (or copper or many other chemical compounds). Suppose that 200 headache tablets, if ingested at one time, would cause death. If this data were plotted on a risk-dose graph, a straight line could be drawn from the data point to the origin. The resulting risk-per-tablet ratio is 0.005, that is, a linear no-threshold theory would predict 0.005 chance of death per tablet. Suppose that in North America 300 million people each take five tablets per year, on average. The linear theory then predicts that 7.5 million people would die each year from taking the headache tablets (300 million people x 5 tablets per person x 0.005 risk of death per tablet). But we know that such deaths do not occur. Clearly, a realistic model would have a threshold. In fact, one tablet a day is thought (and has been advertised) to reduce the risk of heart attack, thus, having a beneficial effect—an example of adaptive response.

The linear no-threshold dose–risk model is useful for regulatory agencies because of its simplicity, and because its application in deriving regulatory limits provides a large safety margin. But at the same time, this model does significantly overestimate risk. It should not be used in most applications: its extreme conservatism leads to extra costs, and can be misleading.

Populations: The Macroscopic View

Let us now step back and look at the forest, rather than the trees.

In Chapter 3 we saw that every part of the earth is radioactive and that humans have evolved with radiation being a feature of their environment. As Charles Darwin stated in his theory of evolution, living things evolve through a process of natural selection and develop defence mechanisms against detrimental stimuli in the environment. It is inconceivable that living organisms would not have evolved and adapted to the levels of radiation encountered in nature—especially since natural radiation was even higher in earlier millennia.

The process of natural adaptability is well illustrated by bacterial diseases and their resistance to antibiotics. Today, we face a potentially serious problem resulting from the genetic versatility of bacteria. Every major disease-causing bacterium now has strains that resist at least one of the over 100 antibiotics used to treat them. Scientists estimate that bacteria typically become resistant to new antibiotics over a period of 5 to 10 years; and it takes at least one decade to develop an antibiotic. So it may be that we are only postponing the inevitable: bacteria will always have the ability to adapt rapidly (on a human scale) to any antibiotics we develop. Mosquitoes and insects are also excellent examples and adjust within a few decades to become immune to DDT and other lethal pesticides that have been engineered specifically to eradicate them.

If bacteria and mosquitoes can adapt to various chemicals, so can humans. Human cells have developed defence mechanisms to cope with radiation. The continued existence of harmful bacteria and mosquitoes are living proof that natural levels of radiation are not harmful to humans.

Locations with High Radiation

Radiation is not constant, but varies from place to place in the world. As noted in Chapter 3, people living in high-altitude cities such as Banff, Mexico City, and Denver receive more cosmic radiation than those living near sea level. High levels of radioactivity also occur in areas with granite outcrops such as Cornwall, England and in the Precambrian shield areas of central and northern Canada. We saw above that people living in areas of higher natural radiation are just as healthy as those living elsewhere. Millions of people have lived their entire lives in Colorado, where they receive more natural radiation (both cosmic and terrestrial) than people living elsewhere in the United States. If low-level radiation was damaging to health then the citizens of Colorado should have greater mortality from cancer than those in other parts of the United States. Studies have shown that exactly the opposite: the cancer rate in Colorado is 35 percent below the national average.

There are many similar examples. A high natural-radiation region in China has radon levels three times higher than average. Han peasants have lived there for 600 generations, yet they show no difference in lung cancer mortality, frequency of hereditary diseases, or any other cancer mortalities compared to the people living in other areas. Kerala, India is underlain by radioactive sand that results in local inhabitants receiving four times the natural radiation dose compared to the rest of India. A survey of 70,000 people over a 35 year period showed no increase in genetic or other adverse health effects.

These observations suggest that the linear no-threshold model should be replaced by a threshold model (curve b in Figure 4-2).

A strong case for the adaptive response model was presented in a comprehensive study of lung cancer mortality and average radon concentration in homes. This study included 1,601 US counties, representing 90 percent of the US population, and concluded that, with or without corrections for smoking prevalence, there is a strong tendency for lung cancer rates to decrease with radon exposure. In spite of extensive efforts, no explanation for this result could be found other than the failure of the linear theory at low radiation levels.

The average natural background radiation (external sources alone) in all US states is 1.3 millisieverts, and in the seven states with the highest natural radiation the average is 2.1 millisieverts. Yet the cancer rates for all malignancies were lower in the seven states with high radiation background (126 versus 150 [per 100,000 per year]) and lung cancer rate was 14.5 versus 20.4 (per 100,000 per year).

Table 4-2 lists a few of the locations in the world where elevated levels of natural radiation occur and shows the doses that the local inhabitants would receive. Note that these dose levels significantly exceed the average natural dose received by Canadians of 2.6 millisieverts per year. If radiation doses of these levels were encountered near nuclear facilities or around nuclear accidents, the media uproar would be sensational and towns would undoubtedly be evacuated. Yet the evidence gathered to date shows that the inhabitants of these areas live normal lives with no signs of adverse effects.

Table 4-2 Locations with High Natural Radioactivity

Location	Annual Dose (mSv/year)
Finland (average)	8
US Rocky Mountain States	6 to 12
Haute Provence, SW France	88
Kerala Coast, India	5 to 70

Occupational Doses

Because radiation is easily detected, it is not difficult to monitor the radiation doses received by workers at nuclear power plants and research institutes. People at such facilities wear personal measuring devices called "dosimeters." Direct-reading and thermo-luminescent dosimeters are the most common.

The direct-reading dosimeter allows the wearer to periodically measure the amount of gamma radiation present at that time. A quartz fibre within the dosimeter measures the radiation and moves along a calibrated scale to provide a reading. Modern electronic dosimeters give instantaneous readings of dose and dose rate.

A thermo-luminescent badge (Figure 4-3) contains a lithium fluoride crystal which records the amount of gamma radiation to which the worker has been exposed over a period of time. When heated, the crystal gives off an amount of light (luminescence) that is proportional to the dose received. This allows automated methods of reading and recording the data from the badges. Compare this to the difficulty of monitoring exposure to non-radioactive contaminants such as might occur, for example, in pesticide factories.

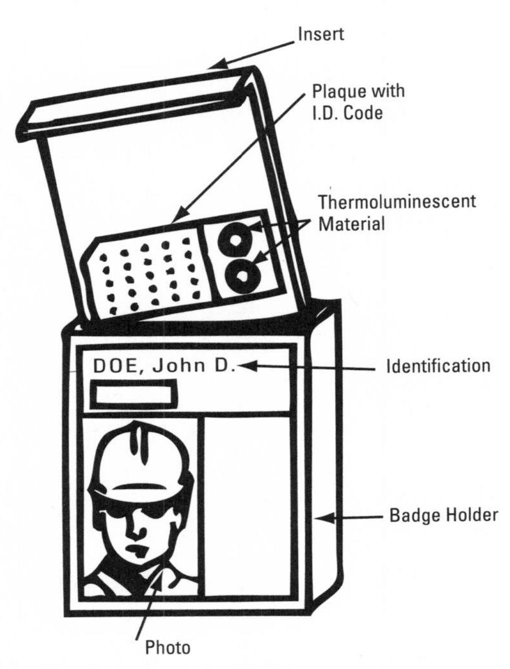

FIGURE 4-3: Thermo-luminescent radiation badge carried by workers at nuclear installations. The thermo-luminescent material is removed periodically and "read" to measure radiation exposure. Usually the badges are also used for identification and carry the person's photograph.

A personal alpha dosimeter is used by uranium miners. Direct-reading and thermo-luminescent dosimeters measure gamma radiation, whereas in a uranium mine the principal risk comes from radon gas, which results in lung exposure to alpha particles. The alpha dosimeter contains a small battery-operated pump that draws air through a special filter, which is sent monthly to a laboratory for counting. More than 500 Canadian uranium miners are monitored by this device.

The Canadian Nuclear Safety Commission sets the exposure limits for what are called Nuclear Energy Workers, which includes employees at nuclear reactors. Whole body radiation exposure must not exceed an annual limit of 20 millisieverts, averaged over five years.

Employees at nuclear reactors are carefully monitored. They wear thermo-luminescent badges when at work and also have regular radiation measurements of urine as well as whole-body counting. In addition, radiation monitors are located throughout the reactor building to detect radioactive contamination. From 1955 to 1987, during more than 133 million worker-hours of operation at Ontario Power Generation nuclear facilities, not a single employee was injured due to radioactivity and there were no medically significant radiation exposures.

A mortality study from 1971 to 1989 of all male workers and pensioners of Ontario Power Generation showed that 34 percent fewer nuclear station workers died during this period than would be expected statistically, based on male mortality rates in Ontario. There were also fewer cancer deaths than would have been predicted in the normal population. These results are due, at least in part, to the close medical attention and study the workers receive.

A 29-year study tracked US Army radiology technicians, who received 500 to 1,000 millisievert doses over a three-year training period in World War II by taking large numbers of X-rays of each other (these are much larger doses than any member of the public would ever receive). There was no difference in cancer mortality among these technicians and a control group of medical technicians.

High lung cancer rates were observed in the 1950s for Canadian uranium miners who worked in areas of high radon concentrations, particularly for the many who also smoked. This observation was a primary reason for establishing nuclear regulation in Canada. The occurrence of cancer was much reduced by improved mine ventilation systems and by non-smoking policies.

The Chernobyl Accident

Chapter 8 discusses the 1986 Chernobyl accident in the Ukraine, the only example ever involving a reactor accident with a large radiation release to the public. At the time of the accident 28 firefighters and reactor operators lost their lives from high radiation doses, and three others died of injuries. There has been considerable debate, however, about the effect of radiation exposures on the public over the longer term.

Huge amounts of radioactive materials were put into the atmosphere—up to 12 trillion becquerels—much of it released in the first ten days of the accident when the reactor was burning. This was roughly 400 times more than the radiation released by the Hiroshima bomb, and between 100 to 1,000 times less than the total radioactive materials released by the nuclear weapons tests of the 1950s and 1960s. Although hundreds of radioisotopes were released, the main radiation exposure to humans and the environment arose from isotopes of cesium, iodine, and strontium. Radioactive materials were deposited in patches (the pattern of deposition varied according to local weather conditions), with some locations far from the reactor having higher levels of radiation than others near it. The radioactivity from the accident was deposited mostly in the Ukraine and its neighbour Belarus, with the largest concentrations near the reactor. Much lower levels of radiation were experienced in other countries in the northern hemisphere. The maximum exposure to an individual outside the then Soviet Union was 0.8 millisieverts, about one-third of one year's natural background exposure.

A special group of policemen, firemen, soldiers, and volunteers were called upon to put out the reactor fire and then to clean up the accident. It is estimated that there were about 200,000 of these so-called liquidators. This number grew to between 600,000 and 800,000 in the years following the accident but many of this additional number received only very low levels of radiation.

The average radiation exposure of the initial 200,000 liquidators was about 100 millisieverts. Approximately 20,000 of them were exposed to 500 millisieverts, and a few hundred received doses in the thousands of millisieverts. As a consequence of the accident, 237 people were admitted to hospital, 134 of them with acute radiation syndrome. Of this group, 28 died within three months of the accident and at least 14 others within about ten years, although it is not clear whether radiation was a factor in all the deaths of this latter group.

All the civilian population living within a 30-kilometre radius of the plant was evacuated, about 116,000 people in all. Pripyat, a city of 45,000 people, was totally abandoned. Of those evacuated, it is estimated that about 10 percent received exposures greater than 50

In Case of Accident Go to the Gas Meter

Potassium iodine or iodate tablets can protect people, particularly children, against thyroid cancer caused by a radiation accident. These pills flood the thyroid with non-radioactive iodine so the gland has no capacity to absorb any additional radioactive iodine. Iodine tablets were provided to some 5.3 million people during the Chernobyl accident, and of these, 1.6 million were children. For this treatment to be most effective the pills must be taken before the radioactive iodine is absorbed.

Ideally, the tablets should be stored in every house, school, and office near reactor installations where they would be available for immediate use in the event of an accident. Like most other types of pills, they deteriorate over time losing their effectiveness and should be replaced periodically. In North America, iodine tablets are generally stored in large quantities at central locations, but they are not replaced regularly.

Some years ago in the United Kingdom iodine tablets were kept up-to-date by gas company employees who entered homes to read the gas meters. They would leave the pills on top of the gas meters where families knew to go in case of a nuclear accident.

millisieverts and less than 5 percent received exposures exceeding 100 millisieverts. Later, an additional 210,000 people were evacuated from high-radiation areas in the Ukraine, Belarus, and Russia out of a population of just over a million.

In the years following the accident, the major radiation-induced illness has been an increase in thyroid cancers, particularly in children. Cancer was caused by inhalation of iodine-131 at the time of the accident or by drinking milk from cows eating contaminated grass. Iodine-131 has a half life of only eight days and was essentially gone in three months. Children born after that period have the normal incidence of thyroid cancer. It is estimated that at least 1,800 additional cases of childhood thyroid cancer occurred in the ten years following the accident. Fortunately, this type of cancer is relatively easy to detect and can be cured by removal of the thyroid; most of these young victims have survived (although the survivors must take thyroid hormones for the rest of their lives). It is estimated that there will also be up to a few thousand adult cases.

If the linear hypothesis were applied to the radiation releases from Chernobyl, about 100,000 deaths would be expected in a population of a few million. Medical teams sent in by numerous agencies have been looking for over 20 years for excess mortality in the population. The people exposed have been carefully monitored, their health has been scrutinized, and their lives placed under a microscope.

In 2005, the Chernobyl Forum—consisting of several United Nations agencies, the World Health Organization, the International Atomic Energy Agency, the World Bank, and the governments of Belarus, Russia, and the Ukraine—issued a report on the eve of the twentieth anniversary of the accident (United Nations, 2005). It stated that 58 people had died to that time consisting of 49 emergency workers and nine children who died of thyroid cancer. The report also projected that the death toll from the accident would eventually reach about 4,000. It concluded that the number of people killed by radiation was much lower than previously thought.

-5-

Electricity and an Impending Energy Crisis

An old man, who had just reached the venerable age of 100, was asked to describe the most important change he had witnessed during his life. Without hesitation he said, "The introduction of electricity. I don't know how we lived without it."

One of our worst nightmares is power failures or brownouts. The ice storm of 1998 in eastern Canada severely disrupted electrical supply when transmission towers and wires snapped under the weight of the ice. Enormous economic devastation ensued, and large segments of the population were stranded without electricity in the middle of winter, some for more than three weeks. The rolling blackouts in California in early 2001 due to lack of electrical-generating capacity also caused considerable stress.

Our society thrives on electricity; it is a fundamental part of our homes, businesses, and factories. When we turn the switch, we expect the light to come on, the kettle to boil, and our shavers, vacuum cleaners, irons, TV sets, computers, washing machines, power tools, and innumerable other items to work. This is all possible because electricity transports energy into our homes and offices. There is hardly a facet of modern society that is not in some way connected to the electrical grid; we are dependent upon a reliable source of electricity.

Yet, in the early years of the new millennium, the ability to supply energy is becoming difficult. We have reached "peak oil," and faced with diminishing supplies, the price of oil in 2008 soared to over $140 per barrel, although the price subsequently subsided with the onset of the major recession. Globally, little potential for hydroelectric projects remains. With human population growing at about 70 million people a year, we are on the threshold of an energy crisis, and providing a reliable energy supply is one of the biggest challenges facing modern society.

Energy and Electricity in Society

Electrical energy use is growing at an enormous rate. This demand is fuelled by two fundamental and irreversible dynamics: the growth in population and the desire for a better standard of living.

In the early days, the human population was small and the world must have seemed an infinitely vast place. But the population increased from a few hundred million to a billion; and then, ever more rapidly, to 6.7 billion in 2008. The globe has become crowded. A further complication is the trend to urbanization. We are living in ever more dense and concentrated conditions. In 1984 only 34 cities had populations exceeding five million; by 2025, the United Nations estimates there will be 93 cities of this size or larger. Enormous quantities of energy are needed to provide necessities such as sanitation, heating, cooling, lighting, and transportation for such large, high-density populations.

As we saw above, early humans used wood as a fuel. Then the burning of coal ushered in the industrial revolution, at the same time as the rapidly expanding population denuded the forests of the United Kingdom and much of Europe. In the past century we have burned oil and natural gas to produce power. Humans have also harnessed renewable energy sources—in particular, falling water, but also the sun, wind, and tides. Following World War II, nuclear energy was introduced as a power source. The progression from wood to coal to hydro/fossil fuels to nuclear represents an evolution from low technology to more complex technologies. As the "simpler" energy resources become depleted, it requires more ingenuity to replace them.

Electrical power is a fundamental part of the aspirations of Third World countries. China, for example, has the ambition to ensure that every household has a refrigerator. Even if they were the newest most energy-efficient models, these refrigerators would require an electrical supply of 20,000 megawatts, or about 25 nuclear reactors of the size of those at Darlington. And this does not take into account stoves, televisions, other appliances or the power needed by the factories to manufacture these goods. Projections indicate that by 2040 the world will need three to four times the electricity capacity that we had in 2000—an incredible increase. It is in the face of this enormous growth in demand for energy that we must consider the role for nuclear power.

The demand for electricity is growing faster than for other power sources as a result of its convenience. Furthermore, battery-driven cars and electrical trains are the transportation methods of the future. The demand for electricity will only increase in both Third World and industrialized countries.

Canada's population is steadily increasing, as is the use of electricity and other energy sources. Because of its cold northerly climate, Canada has a greater need for energy than other countries; other factors influencing the demand for energy in our country are our high standard of living, a vast geography spanning thousands of kilometres, and basic industries relying on processes that use a great deal of energy such as aluminum smelting and pulp and paper manufacturing.

This demand can only be satisfied by energy megaprojects such as the enormous tar sands projects in northern Alberta that produce oil; pipelines thousands of kilometres long that carry oil and natural gas from western Canada to consumers in eastern Canada and the United States; large and complex drilling platforms in the Hibernia oil fields off the coast of Newfoundland; and vast water reservoirs that have inundated millions of hectares in northern Quebec to provide hydroelectricity. Compared to these methods of energy production, nuclear reactors are small and compact.

Canada is fortunate to have a variety of energy sources available in relatively plentiful supply ranging from oil and gas to uranium to waste wood.

Table 5-1 Primary Energy Consumption in Canada in 2006 (from International Energy Agency, Key World Energy Statistics)

By Fuel Type		By Sector	
Oil	40 %	Industrial	38 %
Natural Gas	24 %	Transportation	30 %
Coal	11%	Residential	17 %
Hydroelectricity	10 %	Commercial	14 %
Uranium	9 %	Agriculture	2.5 %
Waste wood etc.	6 %		
Renewables	0.1 %		

Table 5-2 shows the different methods used to produce electrical energy in Canada in 2006. Nuclear accounted for about 16 percent of the total, which compares to the world average of about 19 percent. (For the percentage contributions of nuclear to total electrical energy for other countries see Chapter 7, Table 7-3.)

Table 5-2 Electricity Generation by "Fuel" Type in Canada in 2006

Fuel Type	
Hydro	59 %
Coal	17 %
Uranium	16 %
Oil and Gas	8 %
Wind, Solar, Tidal	0.4 %
Total	591 terawatt-hours

Energy Choices

The choice of which energy sources to use for generating electricity depends on many factors.

Baseload versus Peaking Capacity

Electricity, unlike other manufactured commodities, cannot be stored on a large scale; it must be made—or "generated"—as it is needed. What counts is not inventory, but the capacity to generate. Furthermore, electricity demand is not constant but varies throughout the day, with maximum use in the morning and evening and minimum use during the night. It also varies seasonally. To match this cyclical demand, generating capacity is divided into two categories: base load, which operates all the time, and peaking power, which is turned on and off to meet the cyclical demand.

Because of the complexity of nuclear plants, they are not well suited for turning on and off or for changing power levels frequently. But because of their low fuel costs, they are ideal for providing baseload capacity. Oil-fired plants and natural-gas turbines have expensive fuel but are the cheapest to build and are relatively simple to operate. These are well suited for providing peaking power. In general hydro (water) power can be used across the load spectrum.

Resource Availability

Another consideration is the availability of resources. Ontario, for example, has little indigenous coal, so it must be imported from the United States. This helps to make nuclear cost-competitive with coal. Figure 5-1 shows Ontario Power Generation's Nanticoke Thermal Generating Station, which is fuelled by coal. Alberta has ample low-sulphur coal so nuclear has not been cost competitive there, although in 2008 Alberta announced it will study the feasibility of building nuclear plants. The Maritimes have abundant coal, but this coal contains a high amount of sulphur; once pollution-control devices are added, electricity generated by coal and by nuclear are about the same cost. Pollution-control equipment does not, however, remove the greenhouse gas—carbon dioxide—from coal plant emissions. Currently no commercial technologies are available to capture carbon dioxide and store it permanently.

FIGURE 5-1: The Nanticoke Thermal Generating Station operated by Ontario Power Generation is located on Lake Erie. At full capacity of 1,140 megawatts, the four coal-burning units use more than 4,000 tonnes of coal per day. Note the extensive coal handling facility at the top right.

Only British Columbia, Quebec, Newfoundland (Labrador), and Manitoba have undeveloped hydro resources. In all other provinces, including Ontario, virtually the full potential of hydroelectric power has already been developed. On a worldwide basis, about 60 percent of the hydro power potential is already exploited, and in most industrialized nations the exploitation is close to 100 percent.

(Non-hydro) Renewable Energy

The use of wind power has been growing rapidly in recent years as the technology of windmills advances and costs decrease. From 2003 to 2007, world wind capacity grew at 25 percent each year, and growth is projected to continue at 20 percent for the next five years. In 2008, wind turbines worldwide provided 155,000 megawatts of electricity. That's a lot of turbines—each one has a capacity of about 0.5 to 2 megawatts (although larger ones are in development). On a percentage basis, Denmark is the world's biggest user, with about 20 percent of their electricity produced by wind in 2007. Canada is also investing in wind power with about 1,800 megawatts of installed capacity in 2008; the provincial leaders are Alberta (520 megawatts), Ontario (500 megawatts), and Quebec (420 megawatts). Almost all of this capacity (93 percent) has been installed since 2000.

Wind, this seemingly benign energy source, is not without its complications. Aside from requiring vast areas of land, windmills kill and maim birds of prey such as hawks, falcons, and eagles; these birds soar on air currents near the same ridges that are best suited for windmills.

A more fundamental shortcoming is that the capacities of wind farms are, in practice, much smaller than quoted. Because winds only blow intermittently, a 200-megawatt wind farm generates only about one-quarter the electricity of a 200-megawatt coal or nuclear plant, which run almost all the time. And since electricity must still be supplied to consumers when the wind stops during calms, a wind farm must have backup provided by hydro, coal, or nuclear or its energy must be stored in batteries or by making hydrogen.

Because wind is diffuse, wind farms must be large to make a significant contribution. To replace a 1000-megawatt coal station, for example, about 2000 turbines of the latest design (2 megawatts each) would be required. This would take up somewhere between 11,000 and 20,000 square kilometers of land. Not surprisingly, the NIMBY

("Not In My Back Yard") syndrome will often be a major obstacle to such enormous facilities.

Solar energy can be used to generate electricity using photovoltaic cells. This is commonly done in small-scale applications such as highway signs and remote lighthouses, but it has not yet reached commercial maturity where it contributes significantly to the power grid. Although the potential for solar electricity is good, its costs remain high. Globally, solar had a capacity of about 560 megawatts in 2003, with Japan and Germany leading the way. Solar energy is also well suited for domestic hot water and heating homes and swimming pools. The technology is available, but to date solar water heating is not widely used given its relatively high cost.

Electricity can also be generated by solar-thermal collectors, which are more efficient than photovoltaic cells. The largest solar-thermal facility in the world consists of nine plants in California's Mojave Desert with a total capacity in 2007 of 354 megawatts thermal generated by 936,384 mirrors covering more than 6.5 square kilometres.

Solar energy has similar limitations as wind power; the energy coming from the sun is diffuse and intermittent. To supply China's 1.2 billion people with solar power using today's technology, for example, solar panels would need to cover an area roughly the size of Saskatchewan. Even with significant improvements in solar-panel technology, the area of the collection system would be vast. Although solar energy will make a contribution, China has turned to other sources for the majority of its electrical energy supply, primarily coal-fired generation, which has led to serious air pollution problems. China has also chosen to increase its fleet of nuclear reactors.

Large urban centres with their densely packed apartments, homes, factories, offices, theatres, shopping malls, and subways consume enormous quantities of energy. Furthermore, the world's population continues to increase, cities are growing larger, and the economies of China and India are expanding rapidly. Enormous amounts of energy are needed to supply this growing demand. Renewables such as wind and solar should be used to the maximum extent possible, but due to their diffuse nature and intermittent availability, it is unlikely they will ever make more than a small contribution. Other answers are needed.

Hybrids, Biofuels, Hydrogen, and Fuel Cells

Even if we had abundant electricity, there would still be a large requirement for "transportation" fuels such as gasoline and diesel fuel to power cars, trucks, aircraft, ships, and other vehicles. As these "portable," non-renewable fuels are being depleted, attention is being increasingly focused on developing more fuel-efficient vehicles such as hybrid cars. Also, crops such as corn and sugar cane are being used to make ethanol, known as biofuel. In addition, considerable effort is being directed at developing hydrogen and fuel cells as an alternate power source and as the basis for sustainable energy systems. Electricity can and will play an important role in these advanced methods of powering transportation.

A new brand of "green" cars emerged in 2000 when Toyota launched the Prius and Honda rolled out the Insight and the Civic Hybrid. These hybrid cars combine an electric battery with an internal combustion engine. The Honda Insight burns a mere 3.44 litres of gasoline for every 100 kilometres (68.4 miles per US gallon). The Insight is powered by a 1.0-litre, 3-cylinder gasoline engine that receives additional power from the battery when needed, such as during acceleration from a stop. In turn, the battery is charged from the energy dissipated when the car is braking—a clever use of energy that would otherwise be wasted.

As the price of oil rises, sales are booming and the hybrid car is a commercial success. A potential spinoff is a plug-in hybrid, which is like a regular hybrid but with an extension cord. By increasing the size of the battery and adding the capability to recharge the battery from the grid, the hybrid is effectively converted to an electric car with a small gasoline-engine backup. The car can travel short distances almost entirely on electricity, saving gasoline for longer journeys. General Motors has announced its intention to produce two plug-in hybrids, a Saturn VUE SUV and the Chevy Volt. Toyota has also announced it will have a commercial plug-in hybrid on the road by 2010. With these new hybrid cards on the market, the demand for electricity will only increase.

Ethanol, produced from grains like corn and sugar cane, is being increasingly used as a gasoline substitute. Both Canada and the United

States have initiated steps to ensure that gasoline will contain a fraction of ethanol in the near future. Although ethanol emits less carbon dioxide and pollutants when burned, and is also made from renewable resources, it requires a considerable amount of energy to manufacture. The production of ethanol has had another nasty side effect: it has helped drive up the price of cereal grains. It is an ominous sign when vehicle fuel and human food are locked in head-to-head battle, with the prices of both commodities skyrocketing.

A potentially sustainable future is offered by the use of hydrogen as a fuel. By passing an electric current through water it is possible to separate the water molecule into hydrogen and oxygen gas in a process called electrolysis. (Hydrogen can also be produced by other methods such as steam-methane reforming and the iodine-sulfur process). Hydrogen gas is portable and can be used as a "transportation" fuel. This has been done for experimental cars, buses, and even lawn mowers. When hydrogen burns it produces only heat and water and is therefore non-polluting.

A critical factor here is that electrical energy is required to drive the electrolysis process to make hydrogen. If coal plants were to be used, it would be a step backwards. On the other hand, if hydrogen were to be made from solar electricity and then burned as fuel in vehicles, it would provide an almost pollution-free fuel. At least two small pilot projects of this type have already been tried but were found to be very expensive. Nuclear plants could manufacture hydrogen in large quantities with virtually zero greenhouse gases or air pollution.

Hydrogen is difficult to store and special safety precautions are required in its handling. If hydrogen is to become the fuel of the future, a whole new infrastructure of hydrogen transmission and filling stations will be needed.

Electrolysis can also go in the opposite direction: hydrogen and oxygen can be recombined to produce electricity. This process happens in a fuel cell, which can be thought of as a special type of battery. The fuel-cell concept was originally proposed in 1839, but the first application was as a substitute for conventional chemical batteries in space. The only waste from the fuel cell is water, so it is environmentally clean. Hydrogen batteries are portable and can be used to power vehicles such as cars, trucks, and buses.

So another way of storing energy would be to store hydrogen. For example, excess electricity generated by a solar panel or nuclear reactor could be used for electrolysis. The resulting hydrogen could then be stored and, at a later time, used in a fuel cell to generate electricity.

Hydrogen fuel-cell research and development is currently a very active area. The main challenges are reducing the cost of fuel cells and increasing the electrical power that can be generated. Although still a long way from commercial use, several fuel-cell-powered cars and buses have been tested and demonstrated. In 2008, the first hydrogen fuel-cell car intended for mass production, the FCX Clarity, rolled off a Honda assembly line. Every indication is that fuel cells will at some point in the future replace traditional gasoline engines.

Energy Self-Sufficiency

Another important consideration in selecting an energy source is national energy independence. With oil vital to national economies, and with modern society dependent on oil, few countries want to be at the mercy of supplier countries—especially those that have the potential to become unstable, raise prices, or reduce or withhold exports.

In 2006, the United States spent over $300 billion to import 60 percent of its oil at a price of about $66 per barrel. When the price of oil hovered at $130 per barrel in 2008, the cost of importing oil approximately doubled, adding significantly to the US trade deficit. With oil imports continuing to increase, it is small wonder that the US is quick to get involved when there are crises in major oil-producing regions. Americans have a high standard of living that is dependent on oil, and they are willing to fight to maintain access to this vital resource.

It was the desire to achieve national energy independence that caused France to make a major commitment to nuclear power following the 1973 oil crisis.

Economics

The bottom line in selecting an energy sources is cost, although environmental factors such as greenhouse gas emissions are also starting to play a significant role. Because of its complexity, the capital cost of constructing a nuclear power plant is higher than for fossil-fuel power plants and is also more sensitive to factors such as licensing, construction time, and interest rates. Some of these factors may be influenced by public opinion, which can cause delays and inflated costs that other energy sources may not have to face. It may also influence consumer choice in an open market. The complexity of a nuclear plant also means it must be carefully managed by a large and highly trained work force to avoid expensive downtime. The higher initial capital cost is balanced by lower uranium fuel costs during the operation of the plant.

To decide whether nuclear or fossil-fuel plants are the best choice economically requires assumptions about future interest rates and fuel costs. Therefore, meaningful comparisons can be difficult to make. For example, the year 2000 saw large (60 percent) increases in the price of natural gas in North America. Throughout most of the preceding decade, generating electricity with natural gas was economically attractive. Ontario Power Generation had proposed to convert one of its coal-fired stations to natural gas for environmental reasons, but the escalating gas prices caused this plan to be dropped. It is anticipated that natural gas prices will continue to escalate significantly, following the trend of oil.

In 2008, a new power reactor of the Darlington type would cost around $4 to $5 billion to construct. Sums such as this sound overwhelming and are perhaps best viewed by placing them in perspective with the costs of other energy projects. Newfoundland's Hibernia oil field, for example, has cost billions of dollars to develop, including investment in advanced technology for platforms as well as drilling and extraction methods. The Hibernia oilfield is estimated to contain 525 to 650 million barrels of oil which can produce 3 to 3.5 billion gigajoules of thermal energy. The eight CANDU reactors at Bruce running at 80 percent capacity will generate the equivalent amount of electrical energy in 25 years. The Canadian government has invested $3.5 billion in Hibernia since 1979. The total investment in the nuclear research and development program by the federal government since 1945 has been about $5 billion, of which only a small fraction was for research and development directly related to the Bruce nuclear reactors.

In summary, with expanding populations and economies, the demand for electricity continues to grow and will be further increased with the introduction of electric cars and hydrogen fuel-cells. With supplies of oil diminishing, Canada and other nations will not be able to rely on any one energy source; rather, we will need to use a mix of sources whose choice depends on many factors. Nuclear is one piece of the energy puzzle.

Safety Aspects of Energy

In this section we consider the comparative safety aspects of different energy sources. Detailed discussions of nuclear safety and the impact of nuclear power generation on the environment are discussed in Chapters 8 and 9, respectively.

The public, usually with the Chernobyl accident in mind, views nuclear energy as being particularly dangerous. Detailed assessments of risk demonstrate, surprisingly, that nuclear is actually one of the safer sources of energy, even when reactor accidents are considered.

The potential for severe accidents exists for all major energy systems. Table 5-3 shows deaths caused by different energy sources in the years 1969 to 1996 as compiled by the Paul Scherrer Institute in Switzerland, based on their database of severe accidents (Note 5-1). For the different energy sources, the entire life cycles were included from mining through usage to waste disposal. Hydro-electric power has caused significantly more immediate deaths than any other energy source, per unit of energy generated. This fact—perhaps surprising—is due to the devastating effect of dam failures. For example, some 2,500 people perished in a single dam failure in Macchu, India in 1979.

The main safety concern with coal is in mining, one of the most hazardous occupations in the world, primarily due to methane gas and coal dust, which can explode with devastating results. Coal mining is ten times more dangerous than uranium mining (per unit of energy) in general, and probably 100 times more dangerous than uranium mining in Saskatchewan. Coal mining kills about 100 people per year in the United States, down significantly from about 1,000 per year in the early 1900s. There are also many deaths from occupational exposure to coal dust, namely, the respiratory disease known as "black lung." In China, which produces about one-third of the world's coal, about 6,000 coal miners die each year. Canada has also suffered its share of devastating coal mine explosions. On 9 May 1992, for example, an explosion at the Westray Mine in Nova Scotia claimed 26 lives.

Fatalities from natural gas are generally caused by explosions from gas leaks. A gas pipeline explosion in Guadalajara, Mexico in 1992, for example, killed 200 people. Fatalities in the oil sector are caused by oil platforms capsizing, refinery fires, and fires/explosions during transportation. In 1982, the semi-submersible drilling rig Ocean Ranger capsized and sank on the Grand Banks of Newfoundland, 170 miles east of St. Johns. The entire crew of 84 died. This was Canada's worst oil-related accident.

The Hindenberg

In the same way that every discussion of nuclear power mentions Chernobyl, the Hindenberg disaster is always raised when discussing a future hydrogen economy. The Hindenberg was a German airship, basically a rigid dirigible filled with hydrogen gas and propelled through the air by aircraft engines. Containing 190 million litres of hydrogen gas in its 245 metre length it was the largest object ever to fly. On landing after a regular trans-Atlantic voyage in May 1937, it caught fire, killing 36 people. The film of the disaster is a classic and has been widely shown ever since.

Of the over 4,000 severe energy-related accidents included in Table 5-3 only one was a nuclear accident, although it was certainly a severe and highly publicized one: the Chernobyl accident, which claimed 31 lives at the time. On the basis of immediate deaths, nuclear clearly has the best safety record for accidents of any of the major energy sources.

It should be noted that delayed fatalities, such as late cancer deaths from the Chernobyl nuclear accident and lung cancers from air emissions from coal and oil-fired power plants, were not included in Table 5-3 due to the difficulty in obtaining reliable data. It is recognized, however, that coal has the worst impact in this area. Coal-fired electrical generation is a major contributor to the degradation of Ontario's air quality, which causes more than one thousand deaths per year.

Oil also has significant delayed health impacts and contributes to air quality degradation through air emissions from motor vehicles.

Table 5-3 Immediate Fatalities for Severe Accidents from Different Energy Sources (1969–1996, see Note 5-1)

Energy Option	Fatalities/Unit Energy*
Hydro	883
Coal	342
Natural Gas	85
Oil	418
Nuclear	8

* fatalities per 1,000 gigawatt-years

A Sustainable Environment

More and more, the energy debate is turning away from economics and focusing on environmental impact. There is a growing realization that the rapidly expanding population is placing an enormous stress on the environment, as evidenced by the hole in the atmospheric ozone layer, global warming, air pollution, loss of species, and deforestation, to name just a few of the symptoms. Our voracious demand for energy has been a significant contributor to these problems. There is an urgent need to integrate health and environmental considerations into the growing electrical demand; we must strive to produce electricity in a manner that supports sustainable development.

In the 1970s, global environmental degradation began to be a serious issue. The United Nations Conference on Human Environment held in Stockholm in 1972 established the United Nations Environment Program with the responsibility of building environmental awareness and stewardship. The independent World Commission on Environment and Development was also established with the mandate to look at how development affects the environment.

In 1987, the Commission report, *Our Common Future*, coined the term "sustainable development," which soon drew the world's attention The report offered the following definition: "Sustainable development is development that meets the needs of the present without compromising the ability of future generations to meet their own needs."

Table 5-4 An Environmental Comparison of Coal and Nuclear Power Plants (1,000 megawatts each)

	Land Use (hectares)	Fuel Use (tonnes/year)	Wastes Generated (tonnes/year)	
Coal	70	3,000,000	Ash:	750,000
			CO_2:	7,000,000
			SO_2:	900
			NO_x:	4,500
Nuclear	20	50		900

In 1988, over 50 world leaders supported the report. In 1992, the United Nations Conference on Environment and Development, known as the Earth Summit, was held in Rio de Janeiro and provided enormous publicity and support for sustainable development and produced many initiatives to implement this concept. Sustainable development has since become the cornerstone of many government policies.

Natural gas and oil are relatively scarce commodities that are irreplaceable once consumed. And now, with peak oil signalling that supplies will henceforth diminish even as population and economies continue to expand. A crisis is in the making for industrial society that is built and dependent on this versatile fuel. Oil is the foundation for transportation, industry, agriculture, fishing, and much more. How do we explain to future generations that we ravenously consumed these oil and gas resources?

Of the fossil fuels, only coal—formed from plants that were buried underground and then subjected to intense heat and pressure over millions of years—is found in such great abundance that it could be available for several centuries. Since 1950, the use of coal has more than doubled and in 2007 it accounted for 27 percent of the world's primary energy; coal was used to produce 40 percent of the world's electrical energy and 68 percent of its steel. The downside is that coal is environmentally the "dirtiest" fuel. Research is underway to develop "cleaner" coal plants, including capturing carbon from emissions and placing it deep in appropriate geologic formations for permanent disposal (sequestration). Several decades will pass before such methods are available for routine use.

A 1000-megawatt coal station needs more than three times as much land as a comparable nuclear plant, primarily to store coal. In one year, the coal station burns approximately three million tonnes of coal, compared to 50 tonnes of uranium fuel (60,000 times less). A coal-fired station with modern pollution-control equipment creates over 8,000 times more waste than a nuclear plant, most of which is dispersed into the atmosphere; nuclear waste is carefully contained.

The generation of greenhouse gases that create global warming has become a hot international issue. Table 5-5 summarizes the capacity of four different electrical-energy sources to contribute the greenhouse gas, carbon dioxide, to the atmosphere on a unit-of-energy-produced basis calculated over a full life cycle, including manufacturing, installation, and operation of the facilities. As expected, coal creates the most greenhouse gases per unit of energy. Perhaps surprisingly, nuclear is comparable to the renewables, wind and solar. These require the manufacturing of many units, due to the diffuse nature of solar and wind energy, and this manufacturing process uses energy, which releases carbon dioxide.

Table 5-5 Life-Cycle Emissions of Carbon Dioxide for Electrical Generation Sources

Energy Source	CO_2 (grams/kilowatt-hour)
Coal	about 1000
Solar	60 to 150
Nuclear	6
Wind	3 to 22

Even hydro, a renewable and seemingly benign energy source that creates no air emissions, is not without problems. The construction of dams and reservoirs requires the flooding of large areas of land, displacing many inhabitants and disrupting wildlife and fish habitats as well as historically significant sites. The James Bay project in northern Quebec is a good example. As originally envisaged, the 50-year megaproject was to build 600 dams and dikes, flood an area of 176,000 square kilometres, and displace thousands of indigenous Cree and Inuit. After 20 years and $16 billion, Phase 1 was completed with the flooding of 11,000 square kilometres of boreal forest and tundra. Faced with strong opposition and a surplus of power, the second phase was postponed in 1994, but re-started in 2002. With completion of Phase 2, the total project has a capacity of 15,000 megawatts.

The Aswan hydroelectric dam in Egypt has caused ecological problems. One of these problems is that silt gathers behind the dam instead of being distributed across downriver farm lands. The vast Three Gorges project in China may be another. This hydro-power mega-project will displace millions of people and flood important historic and archaeological sites, therefore facing formidable opposition from local populations.

When dams flood large areas of land, the underlying vegetation decays, producing methane. Since methane is 20 times more damaging as a greenhouse gas than carbon dioxide, some dams that involve the submerging of large masses of trees and plants can cause considerable global warming. The same decay processes release a soluble form of mercury (methyl mercury) from the rocks and soils that is toxic to humans and wildlife.

Although uranium is a finite, non-renewable resource, nuclear reactors can be modified to use other fuels such as thorium (which is more abundant), and also to generate, or "breed," new fuel. Nuclear technology has the potential to be sustainable—that is, to provide an essentially infinite supply of energy (Lightfoot et al., 2006). Although this may seem like a utopian dream, technologies currently exist that allow more nuclear fuel to be created than is consumed; these are described in Chapter 14.

The Role of Nuclear Energy

Nuclear is a different kind of energy; it is not produced by burning a fuel and so does not produce atmospheric emissions. Furthermore, nuclear fuel is very compact: a large amount of energy is contained in a small volume. Under normal operations, nuclear plants are environmentally clean, and their performance under accident conditions (as shown in Table 5-3) is better than that of other energy sources.

For these reasons, some environmentalists who were earlier opposed to nuclear energy have become supporters, e.g., Patrick Moore, the founder of Greenpeace, and James Lovelock, inventor of the Gaia theory. In addition, many organizations including the Club of Rome and the Union of Concerned Scientists support the use of nuclear power to replace the burning of fossil fuels.

As long ago as 1988, the Canadian Parliamentary Standing Committee on Energy, Mines and Resources, recognized that

> The environmental impacts of burning larger amounts of fossil fuel, especially coal, to generate electricity has become alarming. Research is revealing the magnitude of the public health hazard, the enormous economic costs and the environmental destruction resulting from acid gas emissions from fossil-fuelled power plants. The implications of carbon dioxide accumulation in the Earth's atmosphere—an unavoidable accompaniment to fossil fuel combustion—are being studied intensively and the potential for disruptive climatic change is evident. Set against these concerns, the Committee finds nuclear power to be an environmentally appealing technology.

These comments were made by elected representatives, representing all parties, after exhaustive research and testimony, including presentations by anti-nuclear groups.

As we enter the new millennium, we are faced with difficult challenges. The world's growing population desperately needs energy but the era of cheap oil has ended. In addition, the entire biosphere of this planet is suffering and we are faced with global warming, one of the most difficult challenges ever faced by humanity. Nuclear energy, if properly used, can make an important contribution to solving these problems.

-6-

CANDU: The Canadian Reactor

In 1987, the Engineering Institute of Canada ranked the CANDU reactor as one of Canada's top ten engineering achievements of the previous 100 years (other key achievements included the CN Tower and the Alouette space shuttle). The following statement was made about the CANDU reactor as the award was given:

> Perhaps the greatest challenge posed by CANDU was the necessity of adhering to extremely high quality standards in every aspect of the project—in design, manufacture of components, construction and maintenance and operation. This resulted in advances and activity in numerous other related fields. Perhaps best known are the medical applications, such as cobalt therapy machines for cancer treatment. Other spinoffs from the nuclear program include automatic computerized control systems, simulator models, remote handling techniques, and fundamental advances in areas such as metallurgy and chemistry.

The CANDU reactor has clearly had a significant impact on Canadian technology, and is worth a closer look. A total of 22 CANDU power reactors have been constructed in Canada, of which 17 are now operating, 3 are undergoing refurbishment, and 2 have been shut down permanently. CANDU stands for Canada Deuterium Uranium: this reflects the key role of deuterium, or heavy water, to act as a moderator and of natural uranium for fuel. A description of the CANDU design and development is provided by Atomic Energy of Canada Limited (AECL, 1997). Typical CANDU reactors are shown in Figures 2-3 and 2-4.

Reactor Basics

Although nuclear reactors are complex, the basic principles underlying their operation can be understood relatively easily. A nuclear

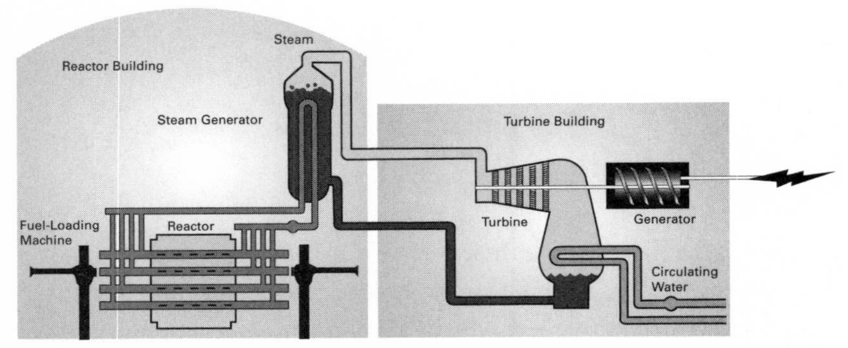

FIGURE 6-1: Schematic of a CANDU reactor. Pressure tubes containing fuel bundles pass horizontally through a vessel called a calandria containing a heavy-water moderator. A nuclear reaction in the fuel bundles heats the heavy-water coolant; this heat is carried to the steam generator where it is transferred to ordinary water in a separate circuit to produce steam. The steam is sent through a turbine, which turns a generator to produce electricity. Circulating water from outside condenses the steam exhausted from the turbine to water, which is then cycled back to the steam generator. The figure also shows the fuelling machines and the reactor containment building.

chain reaction is used to produce heat. The heat turns a fluid into steam, and this steam is used to turn turbines and generate electricity. A schematic of the CANDU reactor illustrating these features is shown in Figure 6-1.

The basic concept is similar to that of a coal or oil-fired power station, except that the energy source is nuclear fuel. The main difference is the way in which the heat is produced. In a nuclear reactor, the heat comes from splitting the uranium-235 nucleus; in a fossil-fuel power plant, the heat comes from burning coal, oil, or natural gas.

Nuclear reactors have the following main components:

- Fuel: a nuclear reactor needs a core of fuel whose nuclei will fission or split when hit by a neutron. This fissile material must be in a sufficiently high concentration that it will sustain an ongoing chain reaction. Virtually all reactors in the world use uranium as the fuel, which in nature is composed of two isotopes, uranium-235 (0.7 percent) and uranium-238 (99.3 percent). A significant complication is that the less abundant of these—uranium-235—is the only isotope that can be efficiently fissioned. As an aside, there are other fissile materials such as plutonium-239 and uranium-233, but these are not currently used to make nuclear fuel in North America (this topic is discussed in Chapter 14).

There are two fundamental ways to achieve a nuclear reaction. One is to increase the concentration of the fissile uranium-235 fuel to the point where a self-sustaining chain reaction is possible, using ordinary water as moderator. This is the approach adopted by most of the reactors in the world—called "light-water reactors"—which use uranium-235 enriched to 3 percent to 5 percent (these reactor types are described in Chapter 7).

The other approach (used in the CANDU) is to use a far more efficient moderator—heavy water—which allows the use of natural uranium (with no need for enrichment) as a fuel. As will be seen, each approach has its advantages and disadvantages.

- Moderator: the neutrons emitted when the uranium-235 nucleus fissions are travelling at such a high velocity that they speed right past other uranium-235 nuclei without splitting them. In other words, the core of fissile fuel must be surrounded by a moderator that slows the neutrons to an appropriate speed. Only a few materials, such as heavy water, graphite, and beryllium are capable of doing this efficiently—that is, slowing the neutrons without absorbing too many of them. Ordinary water can also be used, but the concentration of uranium-235 must be increased to allow for neutron absorption. Moderated reactors are known as "thermal" reactors because fission occurs with slow neutrons. That is, the neutrons are slowed down to have approximately the same energy as the molecules in the moderator, which is related to the moderator's temperature. For this reason these neutrons are known as thermal neutrons.
- Coolant: the enormous amount of heat that is generated in the core must be removed to drive turbines and prevent melting of the fuel and other damage. Fluids that are commonly used include ordinary and heavy water, organic liquids, liquid metals, and gases such as carbon dioxide and helium.
- Control Systems: the rate of nuclear fission must be precisely controlled to maintain the chain reaction at the proper level. Too low a rate of fission, and the chain reaction will terminate; too high a rate and more heat will be generated than the coolant can remove. Typically, rods of stainless steel or alloys containing boron

or cadmium—which are particularly good at absorbing neutrons—are moved up and down inside the core to regulate the reaction. Tubes containing regular water can also be used.

- Safety Systems: because of the large amount of energy that is involved, it is essential that a nuclear reactor have a very reliable and fast-responding shutdown system. One of the most widely used methods to stop the nuclear reaction is to insert rods made of cadmium, a strong neutron absorber, into the core. Alternatively, gadolinium, another strong neutron absorber, can be injected in liquid form into the moderator. As a safety measure, reactors are generally equipped with independent control and safety systems. See Chapter 8 for more information.
- Steam Generator: the heat in the coolant is transferred via a heat exchanger to an independent secondary circuit where it boils and generates steam to drive turbines. Normally, the primary circuit is kept under sufficient pressure to prevent the water from boiling, whereas the water in the secondary circuit is maintained at a lower pressure so it boils. Steam generators—large vessels filled with hundreds of tubes—pass the heat from the coolant circuit to water in the secondary circuit. This system has the advantage that it keeps the radioactivity within the confines of the primary circuit, isolated from the turbines. The steam generators serve as heat exchangers and are sometimes called boilers. It should be noted that the steam generated, in addition to driving turbines, can also be used for heating or other industrial purposes. This was the approach taken at the Bruce nuclear site: the steam was used by heavy-water production plants, right up until they were closed down.
- Turbine/Generator: steam turns a turbine that causes a generator to spin and create electricity. Steam generators and turbine/generators serve the same function (although of different design) as those in conventional power plants fuelled by coal, oil, or natural gas. In both nuclear and non-nuclear power plants about two-thirds of the heat still remains in the steam after it has passed through the turbine. This is removed to condense the steam back to water, and is dissipated as waste heat either to a body of water or to the atmosphere using cooling towers.

CANDU Reactor Design

The CANDU has several unique design features that distinguish it from other international reactors.

- Fuel: the CANDU is fuelled by natural uranium with a concentration of 0.7 percent of the fissile uranium-235. This avoids the complex and costly process of enriching the uranium-235 concentration. Uranium-235 provides the main fissile fuel in a CANDU reactor. However, some of the neutrons striking the nuclei of U-238 do not cause fission but are absorbed and then decay into plutonium-239, which can also undergo fission by neutrons. About 40 percent of the heat energy produced results from plutonium-239 fissions.

  A CANDU fuel bundle is shown in Figure 6-2. The fuel is in the form of small uranium oxide pellets inserted into thin zirconium-niobium alloy tubes, which are sealed with zirconium-niobium caps at each end. Zirconium metal is used to contain the uranium since it absorbs far fewer neutrons than steel, as used in light-water reactors. A Pickering fuel bundle contains 28 elements, whereas a Bruce or Darlington bundle contains 37 elements. A new design of fuel bundle (CANFLEX)—with 43 elements—has been developed that allows more efficient heat removal.

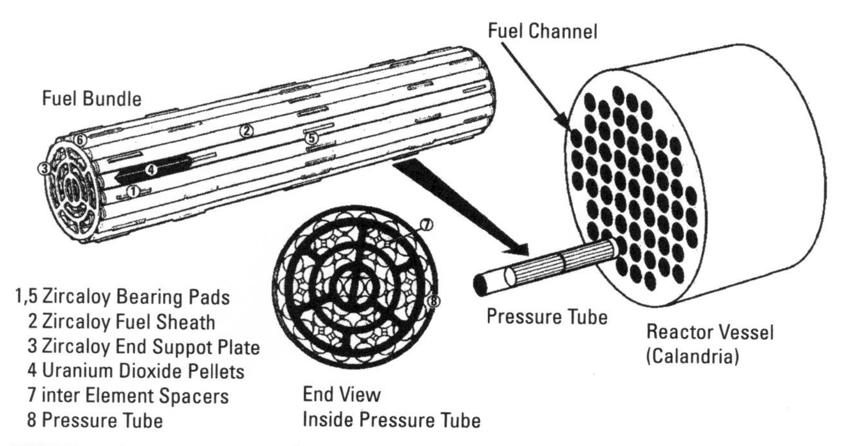

FIGURE 6-2: Schematic of a CANDU fuel bundle consisting of a number of zirconium alloy tubes, or elements, filled with pellets of uranium oxide. The bundles are inserted into horizontal pressure tubes passing through the reactor vessel.

- Moderator: because natural uranium is used in a CANDU, a chain reaction cannot be sustained by using ordinary water as a moderator. Instead, heavy water (that is, deuterium oxide) is used—an amazing 1,700 times more efficient! Since heavy water has almost all the extra neutrons it wants, it slows neutrons in the reactor without significantly absorbing them. When a neutron is emitted, it travels at high speed through the fuel and the fuel cladding, passing into the heavy-water moderator where it is slowed down to "thermal" speed. It then passes into another fuel channel and strikes a uranium-235 nucleus, causing it to split.

  In a CANDU reactor, heavy water also serves as the coolant. The production of heavy water is complex and costly, adding significantly to the initial capital cost of a CANDU, which is greater than it is for light-water reactors. (This higher initial cost is later re-captured by cheaper ongoing fuel costs.) CANDU reactors contain about 0.62 tonnes of heavy water per megawatt, with an annual loss of about 1 percent. Systems are in place to recover heavy-water losses. The latest Advanced CANDU Reactor (see Chapter 14) is designed to use far less heavy water, about 0.21 tonnes per megawatt (AECL, 2007). Projections of the capital costs for the Advanced CANDU Reactor, with significantly less heavy water, are competitive with those of light-water reactors of similar power.
- Reactor Configuration: fuel is arranged in horizontal pressure tubes inside a vessel called a calandria. Because of the lower power density in a CANDU core—a result of the smaller density of fissile atoms in natural uranium fuel combined with the use of heavy water—the reactor core is appreciably larger than it is for light-water reactors of similar capacity. Given this size, the early CANDU designers decided to use a number of pressure tubes rather than a single large pressure vessel to contain the fuel. Although only four pressure tubes are shown in Figure 6-1, there are actually several hundred in a CANDU reactor. The heavy-water coolant flows through these pressure tubes, each of which is enclosed inside a calandria tube. The two tubes are kept apart by spacers called garter springs. This arrangement keeps the coolant and moderator separate. The space between the two tubes is filled with carbon

dioxide gas, which acts as a thermal insulator to limit heat loss from the coolant to the moderator, and is monitored to detect leaks.

The calandria contains several hundred tonnes of heavy-water moderator under low pressure. To ensure a high chemical quality, the moderator is continuously circulated through purification systems. It is kept at a relatively low temperature (approximately 70 °C/160 °F) using a separate heat-exchanger cooling system.

Because the moderator is under constant neutron bombardment, over time some of the deuterium absorbs neutrons to form tritium. As tritium is radioactive, it is useful to remove it from the moderator periodically by a separate purification circuit. There are 380 pressure tubes in a typical CANDU-6 reactor such as at Point Lepreau, New Brunswick, and each tube contains 12 fuel bundles. In total, there are 4,560 bundles in the reactor core. Larger reactors like Darlington have more fuel channels (480) and one more fuel bundle per channel. A bundle stays in the reactor between 6 and 24 months, depending on its location in the core.

- The reactor calandria is contained in a thick-walled concrete and steel structure that provides shielding. In stations built after Pickering A, this shielding structure is filled with hundreds of tonnes of ordinary water and steel balls to provide additional shielding. The CANDU control and safety systems are described in the next section and in Chapter 8.

  In summary, the heat-transport system in a CANDU reactor circulates heavy-water coolant through the fuel channels to remove heat. The coolant, which is kept under pressure to prevent boiling, is then circulated by pumps to a set of steam generators. These act as heat exchangers, transferring heat to a separate water system. The water is maintained at lower pressure so it boils, forming steam which, in turn, drives the turbines to produce electricity. After driving the turbine blades, the steam is condensed and cooled using nearby lake, river, or ocean water and then recirculated.
- On-Power Refuelling: CANDU reactors have on-power refuelling. This is an important feature, allowing them to achieve high capacity factors since they do not need extensive shutdowns every one or two years for refuelling (this is the case for light-water reactors).

W.B. Lewis (1908–1987) was a born in England and was a student of Rutherford at the University of Cambridge. He was invited to come to Chalk River as Director of Research in 1946. One of the first scientists to recognize the potential of nuclear energy, he championed it in Canada and abroad throughout his lifetime. Lewis was the driving force behind the CANDU reactor and is called "the father of the CANDU." He received numerous honours during his life including the Order of Canada.

This feature has resulted in two other unique design elements. First, the fuel channels are horizontal rather than vertical: this allows the fuelling machines to insert fuel bundles at one end of the core while extracting them at the opposite end. A photo of a refuelling machine at a calandria face is shown in Figure 6-3.

Second, fuel bundles can be moved within the fuel channels, usually in two or three stages, to achieve maximum burn-up. Combined with the very efficient moderation provided by heavy water, this results in CANDU reactors using about 25 to 30 percent less natural uranium over a 30-year lifetime than a comparable light-water reactor.

The CANDU pressure-tube design, although offering benefits, also adds complexity. The capital costs of CANDUs are greater than those of their competition because of the use of heavy water. They also involve a lot of "plumbing." As each fuel channel is contained within its own pressure tube, it makes a larger and more complicated core than is the case with light-water reactors, contributing to the higher cost of maintaining CANDU reactors.

The amount of engineering required to develop a CANDU reactor is enormous. The refuelling machines alone are extremely complex, requiring remote operation in extreme temperature and radiation fields. In addition, there are numerous pumps, valves, temperature and radiation sensors, control systems, and many other components.

FIGURE 6-3: Fuelling machine at the face of a CANDU reactor. The machine attaches to the end a fuel channel and adds or removes fuel bundles while the reactor is operating, a unique feature of the CANDU reactor.

CANDU Operation

How do reactor operators control such a large and complex piece of machinery? How do they start up, control, and shut down the nuclear fission process, particularly in a manner that ensures safe operation?

Parameters such as neutron flux, coolant flow, and temperature are measured at a number of key points, usually in duplicate or triplicate. These are monitored by computers and by the reactor operators in the control room. If any measurement is outside its specified range, the power is automatically adjusted either in the particular zone or in the whole core. Four devices are used to control the number of neutrons and hence the power of the reactor: light-water zone controls, control rods, adjuster rods, and moderator "poison."

Light-water zone controllers are tubes running vertically through the core into which regular (or light) water can be introduced. The water absorbs neutrons and lowers the power in that zone. The calandria volume is divided into 14 "zones"; in each of these the power is controlled by a zone controller. Light-water zone controllers will not be used in the Advanced CANDU Reactor.

Control rods are held vertically over the reactor core and can be lowered into the core at varying speeds. These rods contain cadmium, a very good neutron absorber, in a stainless steel sheath. The rods are mechanically controlled by computer but can also be operated manually. Depending on the speed of insertion, they can be used to slowly change reactor power or to quickly shut down the reactor in an emergency. An advantage of the CANDU system is that the shutdown rods are inserted into the calandria against little or no resistance; this is because the moderator is not at high pressure (as is the case for most other reactor types).

Adjuster rods are usually inserted into the reactor core to absorb neutrons, with more absorption in the central region—this keeps the power relatively constant across the core (this is called flux shaping). They can also be withdrawn as fuel gets older to compensate for the build-up of fission products, which absorb neutrons. The adjuster rods can also be withdrawn to allow for short-term inability to refuel. Although usually made of stainless steel, in some reactors the rods contain cobalt-59, which upon irradiation becomes cobalt-60. The cobalt adjuster rods are periodically removed from the reactor and sold to MDS Nordion (see Chapters 11 and 12) where the cobalt-60 is removed and repackaged for medical and industrial uses. About 85 percent of the cobalt-60 used in the world comes from Canadian reactors.

Another important method of ensuring a uniform distribution of neutrons is by arranging the fuel bundles using the online fuelling machines. New fuel bundles can be placed in areas where fuel bundles have been in the reactor longer to ensure a level neutron distribution throughout the reactor core. Continuing adjustments can be made as the fuel is used up.

Another independent shut-down system uses the rapid injection of liquid gadolinium nitrate into the moderator. Gadolinium is a strong neutron absorber. It's worth noting that this method is generally reserved for emergencies: this is because the "poison" later needs to be removed from the moderator by passing it through ion-exchange resins.

CANDU Operating Performance

Until the early 1990s, CANDU reactors had excellent performance records, thanks largely to their unique online refuelling capability. In 1988, for example, CANDU reactors—with gross capacity factors ranging from 84.9 to 90.3 percent—took seven of the top nine places for lifetime performance among all reactors in the world over 500 megawatts in capacity. In 1987, CANDU reactors had probably the best showing ever by any reactor type, holding the top six as well as ninth and twelfth positions amongst world reactors for lifetime capacity. In addition, Pickering Unit 7 established a world record for the longest non-stop operation by any nuclear power reactor of 894 days, ending in 1994.

In August 1983, however, the CANDU reactor suffered a major technical setback. While on full power, a pressure tube ruptured at Pickering Unit 2, spilling mildly radioactive heavy water from the primary circuit into the reactor building. Investigations revealed that the spacer (or garter spring) that separates the pressure and calandria tubes had migrated from its position as a result of vibration, causing the pressure tube to sag and touch the colder calandria tube. This led to embrittlement of the zirconium alloy, and finally a fracture. There was no release of radioactivity to the environment and the reactor was safely shut down by the operators before the emergency safety systems were required.

It was soon discovered that all four units at Station A required retubing. The decision was made to retube Units 1 and 2 at the same time. Unit 2 was out of service for 62 months while all of its pressure tubes were replaced; the retubing of Unit 1 was done in 47 months. The cost was about $400 million, and further costs were incurred in providing replacement energy during these lengthy outages. With the experience gained, the time and cost for retubing Units 3 and 4 decreased significantly, taking 26 and 19 months, respectively.

It was always recognized that CANDU reactors would require retubing at some point during their lifetimes. The very short lifetime of the Pickering Station A pressure tubes (about 12 years) was a cause of great concern. Considerable scientific and engineering effort was and continues to be directed at this problem. Methods have been developed for regularly monitoring the condition of pressure tubes and repositioning garter springs so that retubing should not be required until after about 25 years of operation. In reactors constructed since the early 1990s, new materials (zirconium-niobium alloy with much lower hydrogen content and lower corrosion), as well as a revamped design of the garter spring, ensure that retubing should not be required for about 30 or more years. Moreover, not all tubes will necessarily need replacement.

Research continues to make future retubings more predictable, less frequent, and requiring shorter reactor downtime to conduct them. The new Advanced CANDU reactor, for example, is designed for a 60-year lifetime with a retubing at 30 years; and the retubing process itself will require only one year. This should allow a lifetime capacity factor of over 90 percent.

The Ontario Power Generation Saga

Ontario Power Generation (OPG) has often been in the glare of the political spotlight, but at no time was it more blinding than in the late 1990s. OPG went through a painful and deep-cutting corporate restructuring in the mid-1990s. This was followed by a stunning announcement on 13 August 1997 that seven of its 20 reactors would be laid up over the next year. One other reactor had been shut down earlier due to premature corrosion.

The decision was based on a comprehensive study that concluded that although CANDU technology was sound, OPG had not managed and maintained the reactors well. The downsizing in the mid-1990s and its enormous debt had left OPG with an inadequate workforce and financial resources. For this reason, the seven oldest reactors were laid up so that efforts could be focused on developing improved practices for the remaining reactors. The intention was to bring the laid-up reactors back online in the future.

By 2008, two of the four units that were shut down at Pickering had been refurbished, including retubing at a total cost of $2 billion.

In view of this large expense, it was decided not to bring the remaining two units at Pickering back into service. Two of the older Bruce units were brought back online, but without retubing. An additional two units were being refurbished and retubed at an estimated cost exceeding $3 billion.

Ironically, while OPG was stumbling, reactor performance was improving everywhere else in the world. Driven by competition with coal and gas-fired power, nuclear maintenance has received high priority and has resulted in improved international reactor performance. By 2008, OPG had improved its reactor performance to very good ratings as judged against other world reactors.

So what went wrong in 1997? Many reasons have been put forward. One of the most prevalent is that it had been a mistake to go nuclear in the first place. This is not necessarily the case, as in other jurisdictions such as New Brunswick, Quebec, Argentina, Romania, China, and Korea, CANDU reactors have performed well.

It has also been said that OPG built too much electricity-generation capacity. This was true until about the year 2000, but who was to know in the 1960s and 1970s that a long period of high electrical growth—in which demand doubled every ten years—was to grind to a snail's pace by the early 1990s? Every other province also committed to excess capacity during this period.

The cause, at least in part, appears to lie in OPG having operated as a centralized, government-controlled monopoly. It was not responsible to shareholders and lacked the checks and balances that should have protected its capital investment—the reactors—from deteriorating. Recognizing this problem, Ontario Hydro was split into three separate companies in 1999 and its monopoly over electricity generation in Ontario was ended. The new power-generating entity, OPG, operates in a more competitive marketplace; however, it does still retain control of some 80 percent of electrical generation in Ontario.

Economic Impacts

The Canadian nuclear industry is far larger than most people realize and makes a significant impact on the nation's overall economy. Some economic facts are summarized below:

- Nuclear energy is a $6.6 billion-a-year industry in Canada, generating $1.5 billion in federal and provincial revenues through taxes.
- The Canadian nuclear industry involves 21,000 direct jobs, 10,000 indirect jobs (industry contractors), plus an additional 40,000 spin-off jobs; many of these are high-technology jobs.
- The uranium industry alone accounts for 5,000 of these jobs and is a leading employer of Aboriginal workers.
- Some 150 Canadian industrial firms get all or a substantial portion of their revenues from nuclear business.
- The nuclear industry is responsible for $1.2 billion in exports.
- Canada is the world's leading exporter of uranium, earning more than $1 billion per year.

Many high-technology companies have sprouted from the Canadian nuclear industry, including MDS Nordion, Zircatec Precision Industries (now owned by Cameco), GE Energy, Spar Aerospace, Cameco, Babcock and Wilcox, and CAE Electronics.

Quality assurance and quality control standards used in the nuclear industry are among the highest in any industry. Canadian manufacturers and suppliers to the nuclear program have had to adopt and work to these stringent standards, which have been applied to other product lines and have strengthened the competitive position of Canadian firms in the international arena.

The CANDU Overseas

AECL has been successful in selling the CANDU reactor internationally and has held its own competing against such industrial giants as Westinghouse, General Electric, and France's AREVA.

The CANDU's international involvement began in the early days of Canadian reactor development. In 1955, India and Canada opened discussions that led to Canada supplying an NRX-type 40-megawatts (thermal) research reactor, called CIRUS, to India under the Colombo Plan, a Commonwealth program for support to underdeveloped countries. In the late 1960s, two Douglas-Point type reactors were sold to India, called RAPP-1 (100 megawatts) and RAPP-2 (200 megawatts).

The 137-megawatt KANUPP reactor, modelled after Canada's Nuclear Power Demonstration reactor, was constructed near Karachi, Pakistan in 1972 by General Electric Canada (now GE Canada). It continues to operate today. Two 300-megawatt, Chinese-design reactors have been operating since 2000. KANUPP is the only overseas CANDU reactor not sold by AECL.

In 1974, after RAPP-1 was online but before the completion of RAPP-2, India detonated a nuclear bomb. Canada immediately stopped all nuclear assistance to India. Without any Canadian assistance, India went on to complete RAPP-2 and develop its own indigenous nuclear industry, based on the pressurized heavy-water reactors supplied by Canada.

A CANDU reactor was sold to Argentina and began commercial operation at Cordoba in 1984; it is known as the Embalse reactor. Embalse has maintained a lifetime average of 86 percent capacity factor.

The Indian "CANDUs"

India has an expanding domestic nuclear power program based on a pressure-tube, heavy-water reactor design provided by Canada in the early 1970s. Originating from the RAPP-1 project, they are small compared to typical CANDUs, the largest being 220 megawatts. Twelve of these reactors now generate electricity deployed in one 4-reactor station at Rajasthan and in four 2-reactor stations elsewhere. They are not called "CANDUs" since this is a registered trademark of AECL, and Canada had no part in their construction. Four more of these reactors were commissioned in 2007 and 2008. However, India is also building two pressurized-water reactors of 917-megawatt power. It is interesting to note that the two boiling-water reactors provided to India by the United States in the 1960s, prior to any Canadian transfer of nuclear technology, are still generating electricity today. In 2007, India had 17 operating reactors (3,780 megawatts total) and another 25 planned or under construction (16,336 megawatts total) (CNA, 2008).

Together with Atucha-1 (Argentina's other nuclear power reactor), they supply about 12 percent of the country's electricity requirements.

Korea has made a major commitment to using nuclear power, with a mix of different reactor types, including the CANDU and light-water reactors. Four CANDUs have been purchased from AECL and installed at Wolsong (see Figure 6-4); they came into commercial production in 1983, 1997, 1998, and 1999.

In 2006, Canada's leading private sector companies in the nuclear and power plant field, Babcock & Wilcox Canada, GE Canada, Hitachi Canada, and SNC-Lavalin Nuclear, joined together with AECL to create Team CANDU®. The goal of Team CANDU is to deliver new CANDU power reactors on a turnkey basis in response to the strong demand that has developed for nuclear energy in the new millennium.

In the late 1980s, Romania expressed interest in purchasing four CANDUs to be constructed at Cernavoda. Only Cernavoda-1 proceeded and—as a result of delays in obtaining financing—it was not connected to the electrical grid until 1996. Cernavoda-2 came online in 2007 and the construction of the remaining units is still on hold.

China purchased two CANDU-6 units in 1997. These are located at Qinshan and came online in 2002 and 2003, on budget and ahead of schedule. Also at Qinshan are three pressurized light-water reactors of Chinese design.

In contrast to the lengthy delays at Ontario's Darlington station (where the four reactors were completed in nine to ten years), Korea was able to construct and bring Wolsong-3 into commercial service in a record four years and nine months from award of contract.

-7-
The Global Nuclear Picture

In spite of the controversy that sometimes surrounds nuclear power, it continues to be a key component of today's global energy mix. At the end of 2005, about 15 percent of the world's total electricity was generated by 443 nuclear power reactors located in 31 countries. These reactors have a total electrical generating capacity of approximately 369,500 megawatts. At this time, 27 more reactors were under construction with an additional 21,811 megawatts of capacity.

Light-Water Reactors

As we have seen earlier, there are two main power reactor types in use in the world: those moderated by light (or normal) water and those moderated by heavy water. The CANDU is a pressurized heavy-water reactor (PHWR). The light-water reactors are divided into two types: the boiling-water reactor (BWR) and the pressurized-water reactor (PWR). "Pressurized" means the water in the reactor circuit is kept under pressure so it cannot boil and generate steam. Since steam is necessary to drive the turbines, a PWR requires a steam generator to transfer heat to a secondary circuit where water boils to make steam. In the BWR, steam is produced directly in the reactor core.

Light-water reactors (LWRs) are the most common type in the world. For example, all of the 100 or so operating commercial power reactors in the United States are light-water reactors. In the United States, boiling-water reactors were supplied by General Electric, whereas pressurized-water reactors were supplied by Westinghouse, Combustion Engineering, and Babcock and Wilcox (whose reactor division is now owned by Framatome of France).

The BWR steam cycle employs a single loop as shown in Figure 7-1. In this direct cycle system, the water is heated by the fuel core: it boils and produces steam inside the reactor vessel. The water in a BWR is

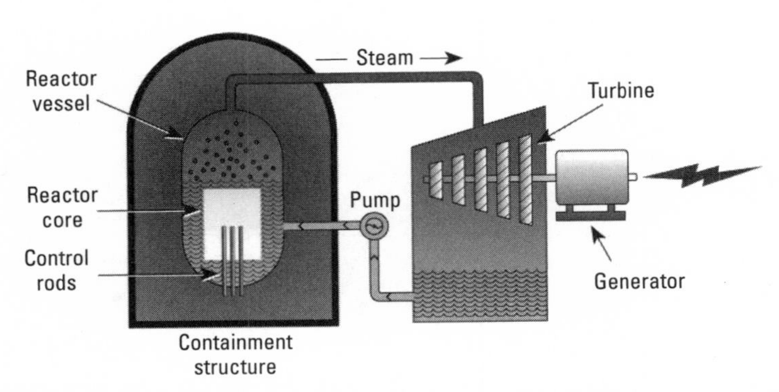

FIGURE 7-1: Schematic of a boiling-water reactor. Steam is generated in the coolant, which passes through the reactor and is sent directly to the turbine to generate electricity. This makes for a simple design but requires careful control of radioactivity in the coolant. Note that the control rods are inserted from the bottom.

actually pressurized (to 7.2 Megapascals) to raise the boiling point of water to 288 °C/550 °F (compared to 100 °C/212 °F at normal atmospheric pressure); the higher temperature improves the efficiency of electricity generation.

The PWR maintains pressure sufficiently high (15.5 Megapascals) to prevent boiling in the reactor vessel. This type of unit employs a two-loop steam cycle, as shown in Figure 7-2. The temperature of the water in the primary loop is about 327 °C (620 °F) as it leaves the core. As in the CANDU reactor, the steam generator transfers heat from the primary to a secondary circuit and produces steam in the secondary circuit for the turbine generator.

Despite the differences between the PWR and BWR, they have many design similarities, resulting from the use of ordinary water as their moderator and coolant. The two are, therefore, grouped together as light-water reactors.

In both the BWR and PWR, the coolant water also serves as the moderator. Because, as we saw above, normal water is not as effective a moderator as heavy water, the uranium fuel must be enriched to about 2 to 5 percent uranium-235. This is a tradeoff. For CANDU reactors, the moderator's isotopic content is modified; in the light-water reactors, the uranium's isotopic content is changed.

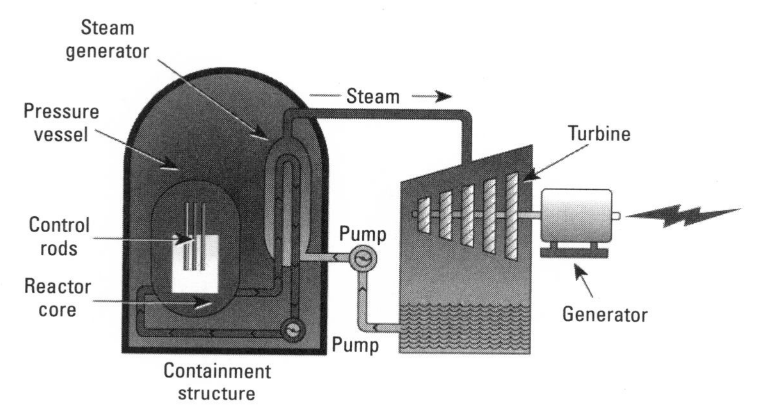

FIGURE 7-2: Schematic of a pressurized-water reactor. As in the CANDU design, the primary coolant loop transports heat from the reactor core to one or more steam generators which produce steam in a secondary circuit. The steam is then taken to turbines to generate electricity.

Although the BWR is conceptually simpler than the PWR in the sense that it does not require steam generators, one issue is that it allows radioactive coolant water to leave the reactor building and enter the turbine generator system. For this reason, the control of contamination and radiation protection for workers is more complicated. The bottom line is that BWRs and PWRs are comparable in terms of total cost.

The fuel assemblies for the two types of light-water reactor are similar, but with some differences. Enriched uranium dioxide is fabricated into ceramic cylindrical fuel pellets (about 0.8 cm diameter and 1.3 cm long). The pellets are placed into long zirconium-alloy cladding tubes to produce fuel pins, also called fuel rods. The fuel pins for a BWR and PWR are similar.

A rectangular array of these pins forms the final fuel assembly. The BWR fuel assembly consists of an array of fuel pins, arranged in a square eight by eight. A fuel channel encloses the fuel-pin array, so that coolant entering at the bottom of the assembly will remain within this boundary as it flows upward between the fuel pins and removes the heat.

The PWR fuel assembly, which consists typically of a 16 by 16 or 17 by 17 array of fuel pins, is about four times larger than that of the BWR. The array is not enclosed by a fuel channel since the behaviour

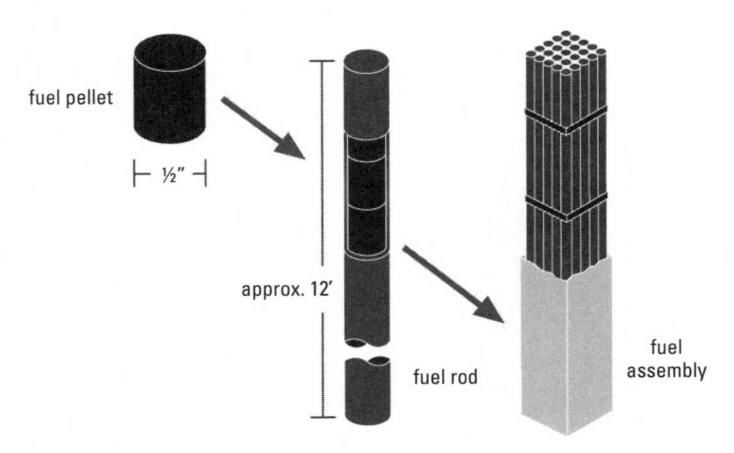

FIGURE 7-3: A generic light-water reactor fuel assembly. Pellets of enriched uranium are placed into long zirconium-alloy tubes. The rods are bundled together in 16 by 16 or 17 by 17 square arrays to form a fuel assembly. Several hundred such assemblies form the reactor core.

of the non-boiling coolant is much more predictable than that of the BWR. Typically, a PWR fuel assembly is about 15 cm x 15 cm in cross section and about 4 metres long. A generic fuel assembly for a light-water reactor is shown in Figure 7-3.

Both the BWR and PWR have vertical reactor vessels. These are large steel structures in which the fuel assemblies, the coolant/moderator water, control rods, and other equipment are contained. In the typical BWR there are 748 fuel assemblies, while there are between 193 and 241 in the typical PWR. The PWR vessel is made of 20-cm-thick steel and is 12 metres high; the BWR reactor vessel is made of 15-cm-thick steel and is typically 21 metres high. (The BWR reactor vessel is higher than that of the PWR because it also contains steam separators and steam dryers that remove water from the steam.) It should be noted that only one manufacturer in the world, in Japan, can make the very large forgings needed for these reactor vessels. This has become a problem recently given the number of LWRs being planned.

In contrast to the CANDU, both the PWR and BWR units must be shut down for refuelling, which takes place approximately every 18 to 24 months. Refuelling alone used to take one to two months but has been reduced considerably in recent years. In 2000, a US record for the shortest refuelling time—just under 16 days—was set by the Braidwood-2 LWR in Illinois. Normally LWR refuelling is combined with a variety of scheduled maintenance tasks and the reactor outages are typically two to three months. In addition, the interval between refuelling shutdowns has been increasing as fuel with a higher enrichment has been used.

Many of the BWR and PWR safety systems are similar to those described for the CANDU reactor. Note that PWR control rods are inserted into the core from above, whereas they are inserted into the core from below for the BWR. In both instances, the rods must be inserted into a high-pressure zone; in CANDU reactors, control rods are inserted into a calandria kept at low pressure. Insertion of control and safety rods into the low-pressure moderator avoids the potential for control-rod ejection: this is a key issue in an LWR design.

The reactor vessels are completely enclosed by a concrete and steel primary containment structure. Both the reactor vessel and the primary containment structures are then contained within a secondary containment structure. In addition, there are numerous other safety systems common to BWRs and PWRs. These systems are intended to prevent or mitigate the consequences of any accident and are based upon "defence by redundancy" as well as "defence in depth" (see Chapter 8).

The principle safety systems used in light-water reactors are collectively known as "Engineered Safety Systems." These are similar to the CANDU safety systems discussed in the next chapter and have been designed, in particular, to deal with "loss-of-coolant accidents"—which are among the worst kind that can happen. These safety systems include:

- reactor trip (shutdown) to provide positive and continued shutdown of the nuclear chain reaction
- emergency core cooling to limit fuel melting
- post-accident heat removal to prevent pressure buildup in the containment
- post-accident radioactivity removal to reduce the radionuclide inventory available for release
- containment integrity to limit radionuclide release.

In recent reactors, most of the safety systems are part of the basic design. In older units, some of the safety systems have been added,

FIGURE 7-4: The Sequoyah nuclear power station operated by the Tennessee Valley Authority. Two 1,148-megawatt pressurized-water reactors are enclosed in the domed containment structures in the centre of the picture. The two large structures are cooling towers where residual heat is cooled by air flowing in the towers.

based on lessons learned from the Three Mile Island accident. Figure 7-4 shows the Sequoyah PWRs operated by the Tennessee Valley Authority in the United States. Note the large cooling towers. Usually the steam leaving the turbine is cooled by water when this can be obtained from a nearby river, lake, or ocean. Where water supply is limited, cooling towers that transfer heat to the air are used.

More recent commercial reactors within the United States are generally of larger capacity than their predecessors.

Although no new reactors have been constructed in the United States since 1990, the first reactor orders in almost 30 years were placed in 2008. In addition, the existing reactors have steadily increased their electrical production. Between 1990 and 2005, US nuclear electricity production from the existing reactors increased by 28 percent. Part of this increase has come from higher power ratings of the reactors and the remainder from improved performance. The average capacity factor for US LWRs was 89.7 percent in 2005.

Table 7-1 summarizes the main differences between the PWR, BWR, and the CANDU reactor. The CANDU core is larger and more complex than the LWR core, and this is because of the pressure tubes. However, this provides some extra safety features as the pressure tubes separate the moderator from the coolant—the moderator is kept at a much lower temperature and would help remove heat in case of an accident. In the LWR, the moderator and the coolant are one and the same and are kept at high temperature.

Table 7-1. Comparison of BWRs, PWRs, and CANDU reactors

	LWRs	**CANDU**
Fuel	Enriched Uranium (2–5% U-235)	Natural Uranium (0.7% U-235)
Fuel Assembly	(BWR) 8 x 8 square array (PWR) 16 x 16 / 17 x 17 square both about 4 m long	Bundle of 37 pins in cylindrical arrangement about 10 cm in diameter, 50 cm long
Number	(BWR) 48 fuel assemblies (PWR) 193 to 241 assemblies	4,560 to 6,240 bundles in core
Reactor Vessel	Large vertical steel tank (pressure vessel)	Calandria (unpressurized horizontal tank) with 380 to 480 fuel channels
Moderator	Ordinary water	Heavy water
Control Rods	Enter into high-pressure, High-temperature zone	Enter into low-pressure, low-temperature zone
On-line Fuelling	No	Yes

Other Power Reactor Types

Although less common, some other reactor types currently in operation are described below.

Gas-Cooled Reactors use graphite as moderator, either natural or enriched uranium as fuel, and carbon dioxide or helium gas as coolant. Most of the early reactors of this type were built in the United Kingdom and are called Magnox reactors (Magnox refers to the magnesium oxide alloy used to sheath the uranium fuel). Britain has developed an advanced gas-cooled reactor. Although carbon dioxide was used in some early commercial reactors, helium has become the gas of choice because it is chemically inert; for this reason it does not easily participate in chemical reactions. The helium is pressurized to about

6.9 Megapascals and its temperature is raised to about 700 °C, which is considerably higher than for light and heavy-water reactors. Most of the Magnox reactors and many of the gas cooled reactors in the United Kingdom have now been decommissioned because of their age.

Light-Water Graphite Reactors use graphite as moderator, and enriched uranium fuel and boiling water as coolant with a pressure tube design. These RBMK reactors are used in Russia and include those at Chernobyl.

International Summary

In 1987, there were 418 power reactors in 26 countries with a total electrical generating capacity of approximately 308,000 megawatts. By 2005, nuclear power had grown by about 17 percent to 443 power reactors in 31 countries with an electrical generating capacity of approximately 369,500 megawatts (see Table 7-2), supplying about 16 percent of the world's electricity. It is clear that nuclear energy is a key component of today's global energy supply; moreover, it is also growing, particularly in developing countries.

Table 7-2 World Nuclear Capacity

Year	1987	2000	2005
No. Countries	26	31	31
No. Reactors	418	433	443
Total Capacity (megawatt)	308,000	349,000	369,500

Table 7-3 shows the number of nuclear reactors (and their capacity) for the nine countries with the largest nuclear power programs. The United States has the world's largest nuclear program, with over 100 reactors; these have a cumulative electrical capacity of 99,000 megawatts. Canada is in ninth position. It has fallen from sixth position following the shutdown of seven of Ontario Power Generation's reactors for extended maintenance.

France is far ahead of all other large countries in terms of the percentage of total nuclear electricity generation—78 percent. The environment has been a big winner, as France reduced annual carbon dioxide emissions to 6.2 tonnes per capita, compared to 10.3 tonnes for Germany and 19.7 tonnes for the United States. The equivalent

Table 7-3 Nuclear Power by Nation (2005)

Rank	Country	No. Reactors	Capacity (megawatts)	Percent Total Electricity
1	USA	104	99,000	20
2	France	59	63,000	78
3	Japan	56	48,000	26
4	Germany	17	20,000	28
5	Russia	31	22,000	16
6	Korea (S.)	20	17,000	37
7	United Kingdom	23	12,000	20
8	Ukraine	15	13,000	48
9	Canada	18	13,000	15

Table 7-4 Reactors Summarized by Type (2005)

Reactor Type	# Units	Net Megawatts
Pressurized (Light) Water Reactors	267	241, 000
Boiling (Light) Water Reactors	94	84,300
Heavy-Water Reactors	41	21,000
Graphite-Moderated Reactors	16	11,400
Gas-cooled Reactors	22	10,800
Other	3	1,000
Total	443	369,500

value for Canada is 17.2 tonnes. Emissions of sulphur and nitrous oxides—leading contributors to acid rain, smog, and respiratory diseases—have been reduced by a factor of more than five.

Table 7-4 summarizes the different reactor types by number of plants and total energy capacity. It is seen that pressurized light-water reactors are the dominant reactor type in the world, comprising about 60 percent of all reactors and 65 percent of total generating capacity. Heavy-water reactors such as the CANDU are in third place, comprising 9.3 percent of reactors and 5.7 percent of nuclear electrical capacity.

In 2008, there were two orders, each for two AP1000 reactors, by utilities in South Carolina and Georgia. These were the first contracts for new reactors in the United States since 1978. The primary reasons for the thirty-year hiatus were flat electricity demand, cost, and the public perception that nuclear reactors are not safe. In the Pacific Rim, however, nuclear construction has taken place and continues to grow. At the beginning of 2006 there were 27 reactors under construction,

France—The Nuclear Sheik of Europe

France, in contrast to neighbouring countries, has adopted a very proactive nuclear energy policy. In 1973, the Arab oil embargo and the huge increase in oil costs left France, which has only limited oil reserves, in a vulnerable position. To avoid a repetition of such a humiliating situation, France embarked on a major program to become energy self-sufficient. With ready access to uranium supplies, France decided to make its electrical supply primarily nuclear. Since 1973, France has constructed 59 nuclear plants, and now not only supplies 78 percent of its own electricity but also exports electricity to England, Germany, Switzerland, and Italy. The latter two countries, which had moratoriums on further nuclear power plant construction, both decided in 2007 to build new reactors. The cost of electricity in France is now one of the lowest in Europe. For the decade up to 1993, average electricity prices declined 20 to 30 percent for industrial users and about 20 percent for residential customers. France is now the largest electricity exporter in the world and has become the nuclear sheik of Europe.

most of them in the rapidly expanding economies of Korea, China, Japan, and Taiwan. In addition, many of the older plants are planning major upgrades so they can obtain life extension permits and operate for many more years, rather than closing.

Reactors That Do not Generate Conventional Electricity

Although the main use of nuclear reactors is to generate electricity for land-based grids, they have other applications as well. There are approximately 280 research and isotope production reactors operating in 54 countries. Some of the research applications of these reactors, which are much smaller than those that generate electricity, are described in Chapter 16.

Nuclear power is particularly well suited for craft that must travel for long periods without refuelling. Currently there are about 250 ships, mostly submarines with some icebreakers and aircraft carriers, powered by more than 400 small nuclear reactors. These ships all use pressurized-water reactors with special fuel to enable them to go years between refuelling. Owing to the end of the cold war, arms reduction, and obsolescence, many nuclear submarines are being decommissioned.

Another type of craft that requires energy for long voyages is the space probe. On October 15, 1997, at Cape Canaveral, Florida, the Cassini space probe was launched on a 3.5-billion-kilometre voyage to explore Saturn and its rings and moons that will last eleven years. Vital to the success of the mission are its nuclear thermoelectric generators that convert decay heat from the radioisotope plutonium-238 into electricity to power the spacecraft and its 12 scientific instruments.

The US National Aeronautic and Space Administration has launched 23 deep-space probes during the past 30 years that used plutonium-238 (which has an 88-year half life). This, unlike its cousin isotope plutonium-239, is not fissile; but it does provide heat from its radioactive decay. Radionuclide energy is necessary because at those distances, the sun is too weak to provide energy and nuclear is the only energy supply that can be used over such extended periods. In 1978, Canada had to clean up the radioactive debris from Russia's Cosmos 954, a nuclear-powered satellite that fell down over the Northwest Territories. Canada has never launched any nuclear ships or satellites.

Apart from marine propulsion, research, and isotope production, nuclear reactors are seldom used for non-electric applications. There

Steam Generator Exports

A steam generator transfers heat from the reactor coolant to a separate circuit where it turns water into steam, which is then used to drive turbines to produce electricity. Like all types of boilers, they tend to corrode over many years of service and must be replaced periodically. Babcock & Wilcox Company of Cambridge, Ontario, has become a leading manufacturer and exporter of steam generators. They supply them not only for CANDU reactors around the world but also as replacement steam generators for US PWR reactors.

are some energy-intensive applications, however, to which they are well suited that we may see in future years, including desalination of sea water in arid regions, extracting and refining oil from the subsurface and from Canada's oilsands, and district heating. There is already some experience with cogeneration—that is, putting the residual heat from electricity generation to beneficial use.

-8-

Safety: The Prime Imperative

The accidents at Three Mile Island and Chernobyl have made most of us aware of the serious ramifications of a nuclear mishap. A nuclear reactor poses a unique hazard that is not associated with fossil-fuel power plants: it has the potential for rapid overheating, which can damage the reactor and release dangerous radioactive materials.

In this chapter we consider the basics of reactor accidents and the physical and institutional mechanisms in place to protect the public and environment from these calamities.

Reactor Safety

Reactor safety is dominated by the three Cs: control, cool, and contain. The first imperative is to maintain control of the nuclear chain reaction at all times and be able to stop it immediately in any kind of an emergency. The second is to always cool the fuel to prevent it from overheating. The third is to contain any radioactivity that might be released in an accident inside the reactor building so that it cannot escape and harm the public or the environment.

The second of these principles is complicated by the fact that the fuel continues to emit heat even after the chain reaction is halted—this is because of the continuing radioactive decay of the fission products. This heat decays away fairly rapidly with time; but, in the absence of any heat removal, it is sufficient to melt the fuel for the first few hours after shutdown. So the fuel must be cooled even after the reactor is shut down.

Let us look at how a reactor accident might happen. Rapid overheating of the fuel could occur if there were, for example, a large pipe rupture, allowing the coolant to be ejected from the system. Heat would rapidly accumulate: this could melt the core, and could also create large quantities of steam that might carry radioactivity to the environment.

The CANDU reactor, like all well-designed reactors, incorporates a number of barriers to prevent the escape of radioactivity in the event of an accident. First, the fuel is in the form of ceramic pellets with a melting point of 2,840°C. The pellets, in turn, are encased in Zircaloy metal. The fuel bundles lie inside a closed heat-transport system, thick concrete shielding surrounds the calandria, and the reactor is enclosed inside a massive concrete containment building. Finally, the reactor is surrounded by a 1-kilometre exclusion zone inside which no homes are permitted. This multi-barrier "defence-in-depth" approach is a fundamental tenet of CANDU reactor safety design (Figure 8-1).

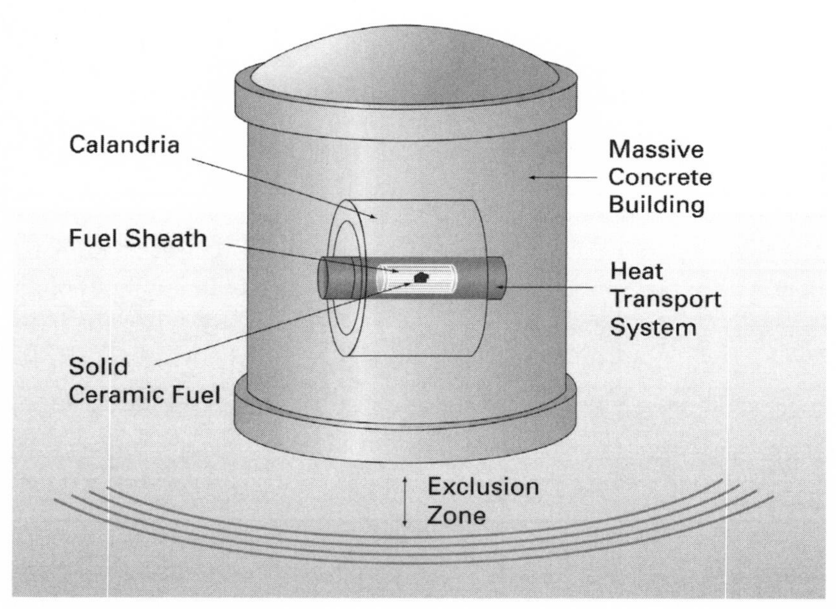

FIGURE 8-1: Schematic showing the "defence-in-depth" safety philosophy used in CANDU reactors. Radioactive materials are contained in ceramic pellets. The pellets are contained in the Zircaloy tubes (fuel sheaths) of the bundles. The bundles are contained in the pressure tubes of the heat transport (cooling) system. The pressure tubes pass through the calandria which is in a massive concrete containment building. The reactor is surrounded by an exclusion zone.

It must be stressed that even in the most spectacular and unlikely type of accident—such as the events of Chernobyl—it is impossible for a reactor to explode like a nuclear bomb. The low concentration of fissile uranium-235 used in reactors and a number of other factors prohibit this from being a physical possibility.

Although the CANDU safety approach emphasizes accident prevention, it also recognizes that failures may occur and that systems and humans are not perfect. The CANDU reactor is designed with a defence-in-depth philosophy to minimize the impact of such failures and imperfections. The Canadian approach emphasizes the separation and independence of three main systems: protective devices, operating systems, and containment.

As discussed in Chapter 6, the reactor's regulating system uses several different devices to control the nuclear chain reaction. In addition, there are totally independent and separate systems for the sole purpose of handling accidents. These are activated only if the regular systems are unable to handle the safe cooling and/or shutdown of the reactor. Among the international reactors, only the CANDU has two independent fast shutdown systems that are physically separate; they are also the only systems to have their own sensors and power supplies, which are separate from the normal control system.

- Shutdown systems: each CANDU reactor has at least two of the following three independent systems that can shut down the nuclear reaction in an emergency.

  First, the moderator can be dumped out of the core into a tank below the reactor. This method was used in the Pickering A CANDUs as well as the reactors preceding them. However, it was too slow. It became obvious that the moderator was a good heat sink for excess heat in accident conditions, and this method is no longer used. Second, cadmium shut-off rods can be inserted vertically into the reactor to absorb neutrons and stop the nuclear reaction. They operate under the action of gravity, assisted by springs; power is not needed for them to function. The third system involves the injection of gadolinium nitrate dissolved in heavy water into the moderator through horizontal perforated pipes. The gadolinium absorbs neutrons and "poisons" the reaction. To ensure redundancy, each of these systems is completely independent of the others and has its own power supply. Furthermore, they are all different *types* of systems (the rods are solid and operate vertically; the "poison" system is liquid and operates via horizontal pipes); they are even serviced by different maintenance crews. The shutdown systems are independent from the control systems (another important feature, as not all reactor designs have completely independent control and safety systems).

- Emergency Core Cooling System (ECCS): should the normal systems that provide core cooling fail, reactors have an emergency core cooling system. Water stored in a reservoir would be injected under pressure over the fuel to remove heat.
- Containment has the dual function of containing the release of radioactivity and also suppressing the pressure surge of any steam release. First, the reactor buildings themselves are constructed of massive steel-reinforced concrete up to 1.8-m thick, as shown in Figure 8-2. In a single-unit reactor (the CANDU 6) the containment building water-dousing systems and air coolers provide short- and long-term pressure suppression. In multi-unit systems, containment is provided by each reactor building. In addition, each reactor is connected to a vacuum building (see Figure 2-3) maintained at a negative pressure. Should there be a rise in pressure from a break in the heat-transport system in any reactor building, valves open and vent the steam and pressure to the vacuum building, where a dousing system condenses and cools the steam.

Reactor Accidents

Since 1952 there have been serious accidents in about ten of the approximately one thousand power and research reactors operated throughout the world (Mosey, 2006). Table 8-1 summarizes these accidents and shows the International Event Scale rating (see Table 8-2).

The first major nuclear accident in the world occurred at Chalk River, Ontario, in December 1952 with the NRX reactor. Due to operator errors, the reactor shutdown rods remained out of the core during a power surge so that the fuel overheated, melted, and released significant radioactivity into the reactor building, as well as causing major damage to the reactor core. The nuclear reaction was stopped by opening a valve and removing some of the moderator. Extensive cleanup was required. No one was injured in the accident and only minor radioactivity was released to the environment, but a culture of safety was thrust onto the

FIGURE 8-2: The 1.8-m thick containment wall of the Unit 1 CANDU reactor in Qinshan, China. The photo was taken just before the concrete was poured. Note the large number of steel reinforcing rods used to make a very strong structure.

Canadian program in its early stages. Fourteen months after this event, the reactor was recommissioned and operated until 1992.

The Three Mile Island accident was caused by a loss of coolant and was further compounded by subsequent operator error. It was initiated by a faulty valve in the cooling system that stayed open when it should have closed. The operators were unaware that this particular valve was open: poorly designed alarm systems in the control room did not provide this information. They then made the significant mistake of manually turning off the Emergency Core Cooling System. As a consequence, fuel in the core was not cooled and a substantial portion of it melted. They only realized their error when radiation alarms in the control room sounded. By the time they turned the Emergency Core Cooling System back on, it was too late. There was substantial fuel melting and core damage.

The radiation was almost all confined to the reactor containment. The public in the immediate area was exposed to only a very small radiation dose, equivalent to a tiny fraction of the yearly natural

Table 8-1 Reactor Accidents

Reactor	Type	Date	Deaths	Event Scale*
NRX (Canada)	Research	1952	None	4
SL-1 (USA)	Military	1960	3	6–7
Fermi-1 (USA)	Prototype	1966	None	1–2
Lucens (Switzerland)	Research	1969	None	3
Browns Ferry (USA)	Power	1975	None	1–2
Three Mile Island-2 (USA)	Power	1979	None	5
St. Laurent-2 (France)	Power	1980	None	4
Chernobyl-4 (USSR)	Power	1986	31	7
Vandellos-1 (Spain)	Power	1989	None	3
Monju (Japan)	Prototype	1995	None	1–2

* See Table 8-2

background dose. Although the reactor suffered significant damage, none of the reactor operators and no member of the public were injured. Nevertheless, a media storm erupted causing considerable psychological trauma to the public in the area. The lack of credible information from the reactor owners and the vacillations of the state government concerning evacuation created a state of fear and anxiety that was totally unwarranted. President Jimmy Carter visited the site; he appointed competent individuals from his government to manage the situation, and generally helped restored public confidence.

President Jimmy Carter

At the time of its accident (1952), NRX was the world's largest research reactor and was being used to test reactor fuel for the US Navy program. For this reason, the cleanup involved both Canadian and US servicemen, one of whom was Jimmy Carter, future president of the United States. Carter was educated as a nuclear engineer and at the time was an officer in the fledgling nuclear arm of the United States Navy. As president in 1979, he also dealt with the Three Mile Island crisis. Thus he had close involvement with the worst nuclear reactor accidents in both Canada and the United States.

FIGURE 8-3: The Chernobyl reactor number 4 (middle with columns) contained in its "sarcophagus" to prevent release of radiation. Reactor number 3, on the left, operated until the end of 2000.

The worst nuclear accident in history occurred at Chernobyl, Ukraine, on April 26, 1986. The operators of a Russian RBMK reactor were conducting an experiment to test a safety system. Inappropriately, the test was conducted at low power, where it was unstable and difficult to control. In the space of a few seconds they lost control and a sudden huge surge in power caused a steam explosion that destroyed the reactor. The moderator, which was made of graphite (a form of carbon) caught fire. Chernobyl had inadequate containment for this size of accident, so large quantities of radioactive materials were released. The moderator continued to burn for many days, sending a plume of radioactive smoke into the atmosphere and creating serious problems for firefighters, several of whom received fatal radiation doses.

The accident took place during the Soviet Union era and was one of the factors that eventually led to the demise of that regime. It was kept secret for some days and only discovered in the Western world during routine radiation monitoring at a Swedish nuclear plant. People in the immediate area of Chernobyl were not informed and were not evacuated for some days. The Soviet authorities, in contrast to the people at the scene dealing with the accident, were confused, ineffectual, and generally incompetent in their management of the accident.

The potential long-term radiological impacts of this accident are discussed in Chapter 4.

It is worth noting that there are at least two major (and many minor) differences between the CANDU and Chernobyl reactors. First, a CANDU reactor does not contain a combustible moderator. The moderator and coolant consist of water and the CANDU reactor structure is made mostly of concrete. By contrast, the Chernobyl reactor used a graphite moderator composed of carbon. A CANDU reactor cannot catch fire and burn as an RBMK reactor can.

The second difference is the quality of the containment. Studies have shown that if the Chernobyl reactor had been equipped with a thick concrete containment such as is standard in CANDU stations (see Figure 8-2), far less radioactivity would have escaped. This does not even take into account the additional protection offered by the CANDU vacuum building.

Reactor accidents such as at AECL's NRX reactor, Three Mile Island, and Chernobyl have shown that serious events almost always have a substantial component of human error. A key factor in the Chernobyl accident, for example, was the lack of a safety culture. A significant CANDU safety feature includes the training of reactor operators to rigorous standards set by the regulatory agency, the Canadian Nuclear Safety Commission (CNSC), which includes the use of sophisticated simulators. The CNSC maintains permanent staff at each of the nuclear stations. In addition, elaborate emergency response plans have been developed that coordinate the activities of municipal, provincial, and federal agencies.

In summary, Canada has been an international leader in developing a systematic approach to minimizing risk in nuclear reactors. This concern with safety was a legacy of the 1952 NRX reactor accident at Chalk River, which formed a wake-up call early in the history of the CANDU. No nuclear plant worker in Canada has lost time from the job as a result of exposure to radiation in over 40 years, and no member of the public has suffered injury or death due to a reactor accident in Canada.

International Nuclear Regulation

Canadian nuclear regulations benefit from the exchange of information and ideas with other countries that have nuclear programs. In particular, Canada participates in a number of international organizations that study the biological and environmental effects of radiation.

The International Commission on Radiological Protection (ICRP) was formed in 1928 to provide recommendations and guidance on radiation protection. Headquartered in London, England, it dealt initially with exposure to medical X-rays and radium. Reorganized and given its present name in 1950, the ICRP has a main Commission consisting of 12 members and a chairman, all scientists from the international community. In addition, there are five standing committees of scientists with expertise in different specialties. Working groups are set up from time to time to deal with special issues and call upon experts from around the world, as necessary. Members of the ICRP are chosen on the basis of their individual merit in medical radiology, health physics, genetics, and other related fields, with regard to an appropriate balance of expertise rather than to nationality.

The ICRP is an influential organization that issues reports on various aspects of the protection of humans against all sources of ionizing radiation. Its recommendations regarding permissible dose limits form the basis of regulations in most countries.

In January 1946, the General Assembly of the United Nations created an Atomic Energy Commission with the purpose of promoting the peaceful use of nuclear energy, setting standards for radiation safety, and developing safeguards against the proliferation of nuclear weapons. The Commission was renamed the International Atomic Energy Agency (IAEA) in 1957.

The major function of the IAEA is to promote cooperation and exchange of technology between its members, with safety being an important factor in both of these activities. One example of its work is the establishment of a system through which a country can rapidly inform others when it has a nuclear accident. As we have seen, this did not happen in the case of Chernobyl.

The International Nuclear Event Scale, which was initiated in 1990 (see Table 8-2) is an integral part of this system. It is also of assistance to journalists and the public in understanding the significance of reactor accidents. A Major Accident is one with major release of radioactive material with widespread health and environmental effects requiring extended countermeasures. A Serious Accident is one with significant release of radioactive material likely to require planned countermeasures. An Anomaly (Level 1) includes, for example, minor problems with safety components but with significant defence-in-depth remaining. All the IAEA countries including Canada have agreed to use the Scale in reporting accidents to each other. As many minor incidents are reported in the media in a sensationalistic and frightening manner, the use of the International Events Scale should help place such incidents in perspective.

Table 8-2 The International Nuclear Event Scale

Level	Public Exposure	Examples
7-Major Accident	Major	Chernobyl
6-Serious Accident	Significant	
5-Accident with Wider Consequences	Limited	Three Mile Island
4-Accident with Local Consequences	Minor	St. Laurent, NRX
3-Serious Incident	Very Small	Vandellos, Lucens
2-Incident	None	Monju, Browns Ferry
1-Anomaly	None	

The United Nations Scientific Committee on the Effects of Atomic Radiation (UNSCEAR) was established in 1955 as a result of concern over radioactive fallout from atomic-bomb testing in the atmosphere. The committee, which consists of 70 to 100 scientists (physicists, biologists, geneticists, medical doctors, and others) from more than 20 countries including Canada, meets once a year. About every five years, UNSCEAR issues a document reviewing and updating scientific information on the levels of radiation to which humans are exposed and the biological effects of this exposure. In 2008, UNSCEAR issued a comprehensive review of the effects of radiation on health all over the world, including the area around Chernobyl.

The World Association of Nuclear Operators, representing nuclear utilities from 33 countries with responsibility for operating over 400 reactors, was established in 1989 in response to the Chernobyl accident. Operating as a voluntary organization rather than a regulatory authority, its objective is to improve nuclear safety through international cooperation and sharing of information amongst its members. It has four regional centres located in Atlanta, Moscow, Paris, and Tokyo.

Each industrialized country also has its own national committee on radiation protection. In the United States, the National Academy of Sciences has established committees on the Biological Effects of Ionizing Radiation (BEIR) that are funded by the Environmental Protection Agency. The committees are staffed by scientists from universities, hospitals, and national laboratories. In Canada this role was filled by the Advisory Committee on Radiological Protection, which reported to the CNSC. This was disbanded in 2001.

Safety in the design and operation of CANDU reactors has continually evolved and improved over the 35-plus years since the first prototype was constructed. Design and operation are constantly reviewed and major research programs are undertaken by AECL, Ontario Power Generation, Bruce Power, Nuclear Safety Solutions, and other organizations. The Canadian program also benefits from the operating experience of nuclear reactors around the world through the Institute of Nuclear Power Operations in Atlanta and the International Atomic Energy Agency in Vienna. As well, the four Canadian utilities and international utilities with CANDU nuclear reactors (Argentina, Korea, Romania, China, India, and Pakistan) share information and conduct joint research through the CANDU Owners Group.

Canadian Nuclear Regulation

Laws governing the use of radioactive materials have been promulgated at both the federal and provincial levels in Canada. The Atomic Energy Control Act (passed in 1946) was the original federal law governing radiation matters. The Act was administered by the Atomic Energy Control Board, the initial nuclear regulator. In 2000, a new act was implemented—the Nuclear Safety and Control Act—establishing a new nuclear regulatory agency: the Canadian Nuclear Safety Commission (CNSC). The CNSC regulates the use of nuclear energy and materials to protect health, safety, security, and the environment, and to respect Canada's international commitments on the peaceful use of nuclear energy.

Responsibility

Today, the CNSC's responsibilities are as wide-ranging as the use of radioactivity. Through a comprehensive licensing process, it regulates all aspects of Canada's nuclear industry. There are two main areas: the nuclear fuel cycle and the use of nuclear substances. These include: power reactors, research reactors, particle accelerators, uranium mine/ mill facilities, uranium refining and fuel fabrication facilities, heavy-water plants, radioactive waste management facilities, and the use and production of nuclear substances and radioactive gauges.

The Commission's duty is to ensure that all nuclear substances in Canada are used in manner that prevents unreasonable risk to human health and the environmental. The Commission is also responsible for nuclear security aspects, transportation, and export of nuclear substances. In addition, the CNSC has some responsibilities regarding insurance against nuclear liability. It contracts for research in areas relevant to its regulations, and in some special circumstances it supervises decontamination projects.

Regulatory Philosophy

The underlying principle governing the CNSC's approach to regulation is to set regulatory requirements incorporating Canadian and international standards and best practices with respect to public and worker risk to radiation exposure. It is then up to the nuclear facility owner (i.e., the licensee) to determine how these requirements will be met. The role of the Commission is to ensure that the licensee lives up to its responsibilities.

Canada and, for example, the United Kingdom share this philosophy of giving the licensee relative freedom in choosing how to meet the regulations. The United States, in contrast, is much more prescriptive and imposes highly detailed regulations. The Canadian approach is called performance-based licensing,· whereas the US approach is referred to as prescriptive licensing. CNSC's approach is felt to allow for greater flexibility in the engineering and design of nuclear facilities, and encourages innovation and continuous improvements, rather than restricting them through specific regulations that apply over the whole life of the facility.

In addition, the CNSC regulates the nuclear industry according to the As Low As Reasonably Achievable (ALARA) principle. That is, actual exposures must not only meet the prescribed limits, but should be kept as far below them as possible, economic and social considerations being taken into account.

As in most countries, Canada's regulations are based on the recommendations of the International Commission on Radiological Protection and on the Nuclear Safety Standards of the IAEA.

Canadian Nuclear Safety Commission Structure

The Commission consists of up to seven people appointed by the federal government. The Chair of the Commission is also the Chief Executive Officer of the CNSC. A staff of about 750 is organized into various directorates spanning the Commission's areas of responsibility.

CNSC's headquarters are in Ottawa and field offices are located throughout Canada and at reactor sites. In addition, the CNSC operates a laboratory in Ottawa that performs radiochemical analyses necessary for their independent evaluations of licensees, as well as maintenance, testing, and calibration of the instruments used by field staff.

The CNSC reports to Parliament through the Minister of Natural Resources Canada. As this is the same Minister to whom AECL reports, some critics feel that there is a conflict of interest, and recommendations are occasionally made to have the CNSC report through an independent Minister, such as the Minister of the Environment.

In setting radiation limits, the CNSC is influenced by the international bodies discussed in the previous section, particularly the International Commission on Radiological Protection, and by other federal departments such as Health Canada.

Licensing Process

A licence must be obtained from the CNSC by anyone who wishes to possess, use, import, or export nuclear materials (uranium, thorium, heavy water, and radionuclides) or by anyone wishing to operate a nuclear facility. Examples of a nuclear facility include a uranium mine and mill, refinery, conversion plant, fuel fabrication plant, heavy-water plant, nuclear reactor, or waste management facility.

The licensing process for all major nuclear facilities such as reactors or waste disposal facilities includes, but is not limited to, three major stages:

1. Site Licence, which requires provincial and federal environmental assessments and public hearings.
2. Construction Licence, which requires a preliminary safety report.
3. Operating Licence, which requires a final safety report.

In addition, there are ongoing inspections and periodic licence renewals. For example, there are CNSC staff assigned full-time at the Pickering, Bruce, Darlington, Gentilly-II, and Point Lepreau nuclear generating stations, as well as at the Chalk River Laboratories. They monitor compliance with licensing requirements and perform routine inspections. They have the authority to order any reactor to reduce power or shut down if licensing conditions are not being met.

Licensees are required to keep detailed records of their operations including radiation exposures of all their employees and concentrations of all radioactive emissions. Worker exposures, measured by devices called dosimeters, must be reported to the National Dose Registry operated by Health Canada. By 2007, the registry contained the cumulative exposure records of over 500,000 individuals who worked in more than 80 different occupations.

At the end of the useful life of a nuclear facility, it must be decommissioned in a manner that is acceptable to the CNSC. This generally involves the restoration and clean-up of the site for unrestricted use, or managing it until it no longer poses a hazard to human health or the environment.

Nuclear Substances and Radioactive Materials

The CNSC issues a large number of licences for use in medical diagnosis and treatment as well as in industry and research. In 1998, there were about 3,700 licences distributed across Canada as shown in Table 8-3.

Transportation

About one million packages of radioactive material are shipped in Canada each year. The CNSC regulates the packaging, preparation for shipment, and receipt of these radioactive materials through the Packaging and Transportation of Nuclear Substances Regulations. In

Table 8-3 Radionuclide Licences

Type of User	No. of Licences
Hospitals and medical	801
Universities and educational	287
Governments	370
Commercial*	2,242
Total	3,700

* includes oil well logging, radiography, gauging, static eliminators, etc.

addition, the CNSC cooperates with Transport Canada in regulating the shipments of radioactive materials under the Transportation of Dangerous Goods Act.

Shipments of radioactive materials can be made only in approved packaging or by special arrangement on a case-by-case basis. To receive approval, packaging must meet established performance standards, which for shipment of highly radioactive materials are very stringent. This type of packaging must withstand a series of tests including heavy impact, fire, and submersion in water. Some packages that have been certified in this way have survived the test of being struck by a locomotive travelling at 165 km/hour.

Accidents in which significant amounts of radioactive material are spilled are very rare. As we've noted above, to date there have been no personal injuries due to radiation exposure in shipments of radioactive materials in Canada.

Nuclear Liability Insurance

The CNSC is also responsible for administering the Nuclear Liability Act of 1976, which established the amount of basic public liability insurance to be maintained by the operator of each nuclear facility. Currently, the operators of nuclear power plants must carry $75 million commercial third-party liability insurance. This coverage is solely to meet claims by third parties in the event of a nuclear accident; it cannot be used to repair the damaged reactor facilities. The Nuclear Insurance Association of Canada, a consortium of insurance companies licensed to do business in Canada, is the only approved source from which reactor owners can obtain liability insurance. The cost to the reactor owner for annual premiums depends on factors such as population density and land values near the reactor.

The Act also provides the federal government the power to establish a commission to deal with situations where the claims could exceed that amount in the case of a nuclear accident. After three decades of reactor operation in Canada, no insurance claims have been paid.

In 2007, the government introduced the Nuclear Liability and Compensation Act in Parliament, which will replace the 1976 Nuclear Liability Act. The Act will raise the liability of nuclear operators to $650 million, allow for review of this amount every five years, increase the time period for submitting claims, and clarify compensation procedures. At the time of writing this book, the bill to approve the act was in its third reading.

Safeguards

Safeguards can be defined as a system of treaties, international agreements, physical measurements, and on-site inspections of nuclear facilities aimed at preventing the misuse of nuclear technology and the diversion of nuclear materials for weapons' purposes. The overriding policy of the Government of Canada is that nuclear materials, equipment, and technology be used for peaceful purposes only.

This policy has evolved in two main stages. In 1974, India exploded a nuclear bomb that used plutonium produced in research facilities supplied by Britain, Canada, and the United States. We noted above how Canada immediately cut off all nuclear technology exchange with India, except for information and assistance involving safety issues related to CANDU technology. Shortly thereafter, Canada began to insist that customer countries provide binding assurances that Canadian nuclear materials, equipment, and technology will not be used to produce nuclear weapons.

In 1976, Canada further required that all customer countries that are non-weapons states must ratify the Treaty on Non-Proliferation of Nuclear Weapons or otherwise accept international safeguards of their entire nuclear program, both present and future. To date, nearly 190 nations have signed the Non-Proliferation Treaty.

The CNSC is responsible for ensuring that Canada adheres to international protocols on nuclear safeguards. These programs help ensure that nuclear fuel is not stolen or otherwise diverted from its intended peaceful uses. The objective is to prevent the manufacture of nuclear weapons. Canada is a signatory of the Non-Proliferation Treaty. The CNSC (jointly with AECL) administers the Canadian Safeguards Support Program, which assists the IAEA in improving methods of ensuring safeguards. This includes improving safeguard equipment installed at the CANDU-6 reactors.

Other Canadian Jurisdictions

Nuclear regulation covers a broad spectrum of activities, many of which relate to areas under provincial jurisdiction, or are within the areas of expertise of other federal departments. To ensure good communication and cooperation with these agencies, the CNSC has developed a joint regulatory process. At the provincial level, for example, the CNSC coordinates with the various ministries responsible for health, labour, environment, and natural resources.

Exposure Limits

The CNSC sets the exposure limits for what are called Nuclear Energy Workers, which includes employees at nuclear reactors. Whole body radiation exposure is not to exceed an annual limit of 20 millisieverts, averaged over five years, with the maximum dose in any given year not to exceed 50 millisieverts. This limit is based on the recommendations of the International Commission on Radiological Protection and on having radiation risks comparable to the risks workers face in other safe industries.

The exposure limit for members of the public is established at one millisievert per year, which is based on the level of risk people are regularly exposed to and is approximately one-third the radiation dose that Canadians receive annually, on average, from natural sources (i.e., rocks, plants, and air). This limit was recently decreased from five millisieverts per year in conformance with the recommendations of the International Commission on Radiological Protection. As described in Chapter 4, there is considerable debate about the effects of low-levels of radiation. In this way, lowering the exposure limit added a degree of safety, if not conservatism. Canadian nuclear reactors and research facilities readily meet this new limit. Operators of CANDU reactors in Canada, for example, set their operational targets at about 1 percent of the regulatory limit, or about 0.01 millisieverts, and in practice achieve even lower levels.

-9-

Nuclear Power and the Environment

With our major cities sitting under pallid umbrellas of smog, the global temperature inching ever higher, and natural spaces disappearing, the issue of environment is on everyone's mind. In this chapter we look at some of the environmental impacts of nuclear power and compare them to those of other energy sources. We are particularly interested in global warming, air pollution, reactor emissions, and low-level nuclear wastes (the issue of high-level wastes is discussed in chapter 10).

Global Warming

Nature spent hundreds of millions of years locking up carbon in underground storehouses of coal, oil, and natural gas. Now our industrial society is emptying this treasure in a century or so—a mere heartbeat in geological time—burning these fossil fuels and releasing enormous quantities of carbon dioxide into the atmosphere. We are fast depleting the earth's store of fossil fuels; once they have vanished, they will not be seen again on earth for at least a billion years.

At the same time, the worldwide loss of forests is reducing the amount of carbon dioxide that is converted to oxygen by photosynthesis. As carbon dioxide levels increase, the atmosphere captures more of the solar infrared radiation that would normally be reflected back into outer space. This so-called "greenhouse effect" drives the temperature of the atmosphere higher.

Concentrations of CO_2 in earth's atmosphere have risen from less than 280 parts per million (ppm) before the industrial revolution to over 383 ppm in 2008, with an attendant increase in the global temperature of about 0.7 °C (2.5 °F). And a minimum of another 0.6 °C (2.4 °F) is committed, as a result of existing fossil-fuel infrastructure. What's more, given the international paralysis in confronting this problem, a rise of 2 °C (4.8 °F) seems inevitable. This number seems small, but in reality will pose very significant problems.

First, global warming is not uniform: temperatures at high latitudes rise more than those nearer the equator. The polar ice caps, for example, are already melting—and they are doing so at rates faster than predicted. This will cause sea levels to rise and will dramatically affect coastlines. For some countries like the Netherlands, it could be utterly devastating.

An alarming consequence of global warming is that it may cause instabilities in the climate system, upsetting the delicate balance between the winds, rains, ocean, and atmosphere. We may see an increase in extremes of temperature and storms, including tornadoes, deadly heat waves, and hurricanes. As mountain glaciers melt, many areas will experience water shortages. And tropical diseases have already started to creep north. Many areas, like Canada's prairies, could be transformed into arid deserts.

Globally, about 25 percent of the carbon dioxide contribution to the atmosphere is caused by electrical generation using fossil fuels. Canada releases the equivalent of 618 million tonnes of CO_2 in greenhouse gases into the atmosphere each year; of this total, electricity generation contributes 17 percent.

When carbon (a major constituent of coal, oil, and natural gas) burns, it reacts with the oxygen in the air to produce CO_2, at the same time releasing energy in the form of heat. Carbon dioxide emission is a fundamental consequence of burning fossil fuels. In fact, for every kilogram of carbon burned, 3.67 kilograms of CO_2 are produced.

Burning natural gas is better than burning coal as it produces about half the CO_2 for the same amount of energy. Natural gas consists largely of methane, however, and methane is some twenty times more effective than CO_2 in trapping heat in the atmosphere. So the various leaks of methane to the atmosphere arising from drilling, transporting, and distributing natural gas reduce its advantage in combustion. Oil combustion is somewhere in between coal and natural gas. A quantitative comparison is given in Table 9-1. Interestingly, firewood is one of the worst fuels in terms of climate change and has the added negative impact that it involves cutting down trees which would otherwise soak up carbon dioxide.

Efforts have been underway to coordinate the international response to this global problem since 1988. The Kyoto Conference of 1997

Table 9-1 Emission of Carbon Dioxide by Fossil Fuels

Fuel	CO_2 per Unit Energy Produced*
Coal	323
Oil	254
Natural Gas	182

* kilogram of CO_2 per megawatt-hour

committed Canada to a 6 percent reduction of six greenhouse gases by the period 2008–12, calculated as an average over those five years. Reductions in the three most important gases (CO_2, methane, nitrogen dioxide) will be measured against the base year of 1990. Reductions in three other gases (hydrofluorocarbons, perfluorocarbons, sulphur hexafluoride) will be measured against 1990 or 1995.

Canada signed the Kyoto Agreement, although the Conservative government in 2007 stated that it plans to have Canada withdraw. Whether Canada is in or out doesn't matter, as it is one of the world's highest per-capita carbon-dioxide emitters and is incapable of meeting the Kyoto targets. In fact, Canada's greenhouse gas emissions, rather than decreasing, increased 32 percent from 1990 to 2005. With oil-related megaprojects like the tar sands booming and the population increasing, this trend can only continue.

Like solar and wind energy and hydroelectricity, nuclear stations do not degrade the environment by emitting greenhouse gases or other smog-producing pollutants. If Canada wishes to make a meaningful contribution to combatting global warming and try to meet the Kyoto goals, nuclear will need to play a role.

Acid Rain and Air Pollution

The burning of fossil fuels, especially coal, produces many harmful air emissions. Sulphur dioxide and nitrogen oxides, for example, enter the atmosphere and cause sulphuric and nitric acid, respectively. Falling as acid rain, these gases have caused extensive damage to lakes and forests in Canada and elsewhere in the world. Hundreds of lakes have been rendered so acidic that they no longer support fish life. Trees are suffering blight in many parts of the world and acid rain is suspected of contributing to the problem. Nitrogen oxides combine with hydrocarbons in the presence of sunlight to cause ground level ozone, one of the most harmful air pollutants. In Europe, about 60 percent of sulphur-dioxide emissions and 35 percent of nitrogen-oxide emissions come from electricity generation.

In addition to sulphur and nitrogen oxides, the combustion of coal releases CO_2, hundreds of volatile organic compounds, carbon monoxide, furans and dioxins (families of toxic chemicals that include chlorine in their molecules), as well as trace amounts of heavy metals such as mercury and radioactive elements into the air. Some of the carbon remains unburned, and is released as fine particles, which are generally less than one five-thousandth of a centimetre in diameter. These particles can float in the air for days with organic compounds and trace elements clinging to them. When inhaled, these particles contribute to acute and chronic respiratory illnesses, including cancer. It is estimated that in Ontario alone air pollution causes 1,900 deaths each year, and coal-burning electrical plants make a significant contribution to this problem.

Each year in the United States alone, air pollutants from burning coal kill thousands of people (estimates range from 5,000 to 200,000), cause at least 50,000 cases of respiratory diseases, and result in several billion dollars of property damage. For example, particulates emitted from US coal-fired power plants are estimated to kill about 30,000 people each year (Freese, 2003).

Introduction of sophisticated pollution-control technologies have significantly reduced some emissions, but such devices are expensive and have increased the cost of coal-produced electricity. Many pollutants escape the pollution-control devices. Furthermore, CO_2, the major global warming gas, cannot be captured by currently available technologies, although research is actively pursuing this goal.

A generally unrecognized fact is that burning coal emits thousands of times more radioactive particles into the atmosphere than do nuclear power plants. This occurs due to the natural radioactivity contained in coal (and all rocks on earth). For example, the United Kingdom's Central Electricity Generating Board estimated that it releases 300 kilograms of uranium into the environment every day from burning coal.

In 1998, Ontario substantially increased power generation from coal-fired stations to replace the electricity no longer available from the seven nuclear stations that were shut down for extended maintenance. Ontario then became one of the leading sources of air pollution in North America due to the chemicals and smog from these plants. Health authorities in Toronto estimated that several hundred additional deaths annually could be attributed to this degradation of air quality. This trend was reversed when the Lakeview coal-fired generating station was closed in 2005 and as some of the nuclear stations came back online.

Reactor Emissions

Nuclear generating stations release small quantities of radioactivity into the atmosphere and adjoining water bodies. These emissions are monitored to ensure they are kept well below levels that could cause harm to humans and the environment. The magnitude of the doses that members of the public may receive due to these emissions are so low they can not be measured directly to see if they are in regulatory compliance. Instead, the Canadian Nuclear Safety Commission has established Derived Emission Limits (DELs) by considering the exposure pathways through the environment by which radioactive emissions could reach members of the public. If emissions are kept below the specified DELs, calculations show that the dose to the most exposed member of the public will be less than the regulatory limit, that is, the whole body dose will not exceed 1 mSv in any year.

Airborne emissions include tritium, argon, iodine-131, and particulates. Water emissions include tritium and other radioactive elements. The nuclear utilities and other provincial and federal agencies monitor all the possible paths by which radioactivity can enter the environment and the food chain in the vicinity of the reactors, including air, precipitation, milk, drinking water, vegetation, algae, and fish. Because radiation is easy to detect with appropriate instruments (see Appendix A), these reactor emissions can be tracked accurately.

The CNSC released data for the period 1988–97 showing that the emissions from all Canadian nuclear power reactors were well below the mandated DELs. In fact, no emissions exceeded 1 percent of those values, a target that nuclear utilities have voluntarily set themselves.

During the more than 40 years of nuclear electricity generation in Canada, which involves 25 reactors, there has never been a known injury to the public involving radioactivity. Furthermore, there has never been a release of radioactivity from any nuclear generating station that resulted in a measurable radiation exposure to any member of the public.

Overview of Nuclear Wastes

In Canada, nuclear wastes are divided into three main categories:

- spent fuel (also known as high-level waste),
- low-level waste, and
- historic wastes.

Spent fuel, or used fuel, consists of fuel bundles which have been used in a nuclear reactor until they no longer produce power effectively. Although these bundles appear the same physically as before going into the reactor, spent fuel is highly radioactive, generates heat, and must be handled by remote control.

There are many definitions of low-level radioactive waste; however, the broadest and simplest is that it consists of all those radioactive wastes that are not high-level waste (spent fuel). Low-level radioactive waste includes a broad spectrum of chemical and physical forms such as filters and resins from cleaning a reactor's coolant, worn-out reactor components, protective clothing, and much more. Although low-level radioactive waste contains many different radionuclides, its radioactivity is relatively small, contributing less than 1 percent of the radioactivity at nuclear reactor stations; the other 99 percent is in the spent fuel. This kind of waste also includes radioactive wastes from nuclear medicine, research, and industrial applications of radiation.

Historical wastes are a special category; here, the original generator is no longer in business or cannot be held responsible. These consist primarily of wastes in and near the town of Port Hope, Ontario, that accumulated between 1933 and 1988 from the uranium refinery and conversion plant operated by the former Eldorado Nuclear Limited. Radioactive contamination, primarily of soils, also happened at a few other locations. The contamination occurred largely because the risks of radioactivity were not well understood in those days.

To provide perspective to the management of nuclear wastes, it is helpful to view them in the broader context of how society manages non-nuclear wastes.

Toxicity

Every home in Canada generates wastes, which are hauled weekly from our curbside. In addition to items such as chicken bones and empty tooth-paste containers, they also contain hazardous materials such as used motor oil, batteries, oven cleaners, paint thinners, pesticides, tires, wood preservatives, drain cleaners, drugs, and much more. Approximately 1.9 kilograms of hazardous waste is produced per person per year, much of which winds up in municipal landfills. In addition, industries produce large quantities of hazardous waste.

Hazardous waste includes, for example, chlorinated organic compounds such as vinyl chloride (an industrial chemical used to make polyvinyl chloride, a common plastic) and trichloroethylene (commonly used as a solvent and degreaser) as well as heavy metals including copper, mercury and lead. Vinyl chloride can cause liver cancer and neurological disorders, and lead can damage the nervous and reproductive systems. These wastes can cause damage to biological systems that is as or more severe than the effects of nuclear wastes. In particular, there are more than 1,500 (non-radioactive) substances that are known to cause cancer.

The toxicity of nuclear waste has been compared to that of other substances by Cohen (1990), who calculated the amounts of reprocessed US nuclear waste that would need to be ingested by a human to be lethal. The results are shown in Table 9-2. For comparison, Table 9-3 shows the amounts of various chemical compounds that would need to be ingested to be lethal.

Comparing Tables 9-2 and 9-3, it is clear that the toxicity of nuclear waste is not significantly different from that of other compounds that are used in industry. If the decrease in toxicity with time is taken into consideration, nuclear wastes are actually less toxic in the long term than many non-nuclear compounds, which retain their toxicity forever.

Table 9-2 Lethal Doses for Ingestion of Nuclear Waste

Years After Burial	Amount Causing Death (grams)
1	0.3
100	2.8
600	28.0
20,000	454.0

Table 9-3 Lethal Doses of Chemical Compounds

Compound	Lethal Dose (grams)
selenium compounds	0.3
potassium cyanide	0.6
arsenic trioxide	2.8
copper	20.0

Nuclear wastes are easier to deal with than non-nuclear toxic substances since nuclear wastes are kept isolated and contained, become less toxic with time, and are easy to detect. In contrast, many non-nuclear toxic compounds such as arsenic trioxide, a herbicide and insecticide, are scattered intentionally on the ground where food is growing, and are even sprayed directly on fruits and vegetables.

Quantities

Let us compare the quantities of nuclear and non-nuclear wastes. The Keele Valley landfill in Toronto, which closed in 2002, is Canada's largest municipal landfill and contains 25 million tonnes of municipal waste. Studies have shown that between approximately 0.1 and 3 percent of a municipal landfill consists of hazardous wastes, which would yield approximately 0.02 to 0.6 million tonnes of hazardous waste in the Keele Valley landfill. This is similar to the capacity of the proposed high-level nuclear waste vault (about 0.25 million tonnes of used fuel). A single nuclear repository would contain all of Canada's nuclear waste, whereas Keele Valley is but one of thousands of municipal landfills in Canada.

The situation is the same for landfills containing hazardous waste (rather than municipal waste). For example, the Clean Harbor Canada (formerly Safety Kleen) hazardous-waste landfill near Sarnia, Ontario, will contain about 7.5 million tonnes of hazardous waste when it is full. This is approximately 30 times more waste by weight than will be contained in the proposed spent-fuel repository. The Clean Harbor facility is one of three hazardous-waste disposal facilities in Canada.

It is clear that society produces much more non-nuclear waste than nuclear waste of similar toxicity.

Duration of Toxicity

Some components of nuclear waste last a long time; for example, plutonium-239 has a half-life of about 24,100 years. Many people are concerned that scientists and engineers cannot build a disposal facility that would retain its integrity for this length of time, which exceeds the period of recorded history. The disposal of non-nuclear wastes, however, involves toxic time spans of even longer duration.

Non-nuclear wastes are composed of organic and inorganic components. The former decompose into innocuous forms such as carbon dioxide, methane, and water over a period of decades to a few centuries. In contrast, inorganic materials such as heavy metals do not decay at all. In effect, they have infinite half-lives. Thus, both non-radioactive and radioactive wastes have a component that decays relatively quickly, namely organic wastes and fission products.

And both wastes have a long-term component, namely inorganic compounds and actinide elements (radioactive elements with high atomic numbers and long half-lives), respectively. The inorganic component of non-nuclear wastes remains toxic forever; the actinide portion of nuclear wastes, such as plutonium-239, decay with time, albeit very slowly. Contrary to what many people believe, the long lifetimes of nuclear wastes are not unique.

Disposal Methods

So how are non-nuclear wastes of infinitely long hazardous life-time disposed? Municipal, industrial, and hazardous wastes are all disposed of in the same manner: they are placed into landfills that consist of large mounds of wastes at ground surface. Modern landfills have plastic and/or clay liners underneath to prevent the escape of leachate, covers on top to prevent the infiltration of rain, and pipes within to collect leachate and gas emissions.

Landfills require large land areas usually near urban centres (Keele Valley landfill, for example, is 380 hectares). Exposed to erosion and precipitation, they must be maintained for long periods of time during which they will emit large quantities of greenhouse and toxic gases. Eventually, no matter what technology is used, liners will break down and leak toxic materials into the groundwater.

In summary, near-surface municipal and hazardous waste landfills are the accepted method of disposing of non-nuclear wastes, although these wastes are of similar toxicity and lifetimes as nuclear wastes. Now let us see how nuclear wastes are managed.

Low-Level Radioactive Waste

Low-level waste is generated by nuclear utilities, medical institutes, and industry, as well as by research organizations and universities. Nuclear utilities generate the following low-level radioactive waste:

- process waste consists of materials, many of which have complicated names—spent ion-exchange resins, filter sludges, evaporator bottoms—that arise from cleaning a reactor's moderator, coolant, and storage-pool water. Radionuclides enter into the coolant and moderator by leakage from fuel and by nuclear activation of corrosion products. Typical of the latter are cobalt-60, iron-55, iron-59, and manganese-54.
- trash including mops, contaminated equipment, used protective clothing, temporary floor coverings, which largely arise from housekeeping operations in those parts of a reactor where radioactivity is present. This category has the largest volume and the lowest activity.
- irradiated components such as control rods, which have reached the end of their useful life or have been damaged.

Typically, a single CANDU reactor in one year produces 12.5 cubic metres of process waste, 250 cubic metres of trash, and 25 cubic metres of irradiated components.

Low-level wastes at hospitals and universities arise from the practice of nuclear medicine and from biomedical and other research and include spent technetium-99 generators, expired vials of radiopharmaceuticals, and contaminated syringes, glassware, gloves, and absorbent pads. Only a few of the nuclides commonly used in nuclear medicine have half lives longer than 28 days; these include iodine-125, cobalt-57, and ytterbium-169, which are used in very small quantities.

All radionuclides employed in nuclear medicine (see Chapter 11) are prepared and processed by radiopharmaceutical manufacturers. The production of some of them, in particular molybdenum-99 and iodine-131, results in relatively large amounts of radioactive waste.

Industry generates a variety of low-level waste including cobalt-60, cesium-137, americium-241, and other radioactive sources that have reached the end of their useful lives. These sources as used in sterilization, radiography, geophysics, and a host of other applications (see Chapter 12).

There are a number of practices used to manage low-level waste including holding them in storage until they have decayed to innocuous levels (used for radionuclides with short half-lives), incineration, compaction (where materials are squeezed into a smaller volume by powerful presses), various liquid treatments such as ultrafiltration and reverse osmosis, and disposal. Extensive research has been conducted to develop and improve these methods, so they can deal with the large spectrum of chemical and physical forms of nuclear waste.

In Canada, no disposal of low-level waste has yet been carried out, where disposal means permanent entombment with no requirement for future maintenance. Instead, long-term, managed storage is used. Internationally, disposal methods include burial in shallow trenches (United States), mined-cavern disposal (Sweden, Finland, United States), and near-surface concrete bunkers (France).

In Canada, low-level wastes are treated and stored at two main sites: the Western Waste Management Facility Site of Ontario Power Generation and Chalk River Laboratories of AECL.

The Western Waste Management Facility

Located at the Bruce Nuclear Power Development Site, the Western Waste Management Facility is owned and operated by Ontario Power

Generation (OPG). It treats and stores low-level radioactive waste from all 12 of the reactors operated by OPG as well as the eight reactors operated by Bruce Power at this location. The 8-hectare site (see Figure 9-1) includes an incinerator with a nominal capacity of between 200 and 250 cubic metres per month, a mechanical compactor with waste placed into 205-litre steel drums, and a baler.

FIGURE 9-1: Aerial view of the Western Waste Management Facility at the Bruce nuclear site, Ontario (Courtesy Ontario Power Generation).

All waste at the Bruce site is stored in a solid, retrievable manner in engineered facilities having a 50-year design life. There are four types of storage facilities:

- Reinforced concrete trenches are used for the trash component of low-level radioactive waste.
- Cylindrical holes are used for process wastes. They are typically 0.7 metres in diameter and 3.5 metres deep and are constructed of concrete with a removable steel liner.
- Eleven storage buildings are used to store wastes with low radiation fields (less than 10 mSv per hour). Wastes originally in trenches are transferred to the buildings once their radioactivity has decayed sufficiently. Figure 9-2 shows the storage casks inside one of these

buildings. Two other buildings store intermediate-level wastes arising from refurbishment of Bruce reactors.

- Quadricells. These are double-walled, above-ground, reinforced-concrete structures used for storing the most radioactive low-level waste.

Each year approximately 3,000 cubic metres of low-level waste and about 290 intermediate-level waste are placed into storage. The entire storage area is underlain by an engineered drainage system that is monitored.

FIGURE 9-2: Interior view of a storage building at the Western Waste Management Facility (Courtesy Ontario Power Generation).

As the Western Waste Management Facility is for storage only, those wastes that have not decayed to innocuous levels will need to be transferred to a disposal site at some point in the future. Ontario Power Generation is proposing a repository with a capacity of 160,000 cubic metres under the Bruce site in sedimentary rock strata, which will contain all the low-level waste stored in the Western Waste Management Facility as well the low-level waste generated from all 20 Ontario reactors during their lifetimes. The repository will consist of a number of rooms on one horizontal level at a nominal depth of 500 to 700 metres. The site is undergoing extensive investigation, the environmental assessment process has begun, and, assuming appropriate licences are received, the repository is scheduled to start receiving waste in about 2017.

Chalk River Laboratory

Atomic Energy of Canada Limited has been managing low-level waste from its own research operations and from other waste generators in Canada for over 50 years and has also conducted research into classifying, treating and disposing of wastes. The bulk of this work takes place at its Chalk River Laboratory; some was formerly done at its Whiteshell Laboratory. AECL receives and manages the low-level waste from nuclear medicine, industry, and research for all of Canada. This service is essential to these organizations.

Activities at Chalk River are focused around the Waste Treatment Centre where liquid wastes are treated by passing them through an evaporator that concentrates the waste. An ultrafiltration unit is also available. The final product is immobilized in bitumen. Solid wastes are classified and treated by compaction and baling.

The Chalk River site has extensive storage sites which have evolved over a long period. These consist of several waste management areas where low-level waste is stored in pits, unlined and asphalt-lined trenches, concrete structures, and a waste tank farm. One area stores historic wastes, mostly contaminated soils, on behalf of the Low-Level Radioactive Waste Management Office.

Current practice is to segregate the wastes by physical and radiologial properties, reduce their volumes and place them into in-ground concrete bunkers or buildings called shielded modular above ground storage (SMAGS). The first of six SMAGS has been completed. Five more will be built at intervals of three to four years providing storage capacity for the next 20 to 30 years.

Two modular above ground storage buildings (MAGS), similar to SMAGS but smaller and not shielded, were constructed in 2004 along with a supercompactor. These modular structures store low radioactivity materials in steel containers that were previously placed in sand trenches.

Considerable effort is focused on minimizing waste volumes. For example, a waste assessment facility was built in 2007 that monitors wastes so that non-radioactive items can be removed and recycled or disposed of as regular waste; this reduces the amount of low-level waste.

An intermediate-depth mined cavern, called the Geological Waste Management Facility, is proposed for low-level wastes with a hazardous life greater than 500 years. The first borehole to explore the suitability of the Chalk River site was drilled in 2007. If conditions are appropriate and regulatory approval is received, the target date for commencement of operations is 2025.

Historical Wastes

Historical wastes are found primarily in the Port Hope, Ontario, area where Eldorado Nuclear (now Cameco Corporation) initially refined ore to obtain radium for medical purposes and luminous paint. In later years, radium refining was replaced by uranium refining. Wastes from these operations contain the same radioactive elements as in the original ore, although the concentrations and chemical forms may be different. Historical wastes are the responsibility of the Low-Level Radioactive Waste Management Office.

Port Hope, Ontario

Between 1948 and 1955 wastes were deposited in the Welcome waste management facility located about one kilometre west of Port Hope. The facility contains about 12,000 cubic metres of process waste and 255,000 cubic metres of contaminated soils. A treatment facility collects and treats contaminated ground and surface water.

The Port Granby waste facility, located on the shore of Lake Ontario 16 kilometres west of Port Hope, received wastes between 1955 and 1988. It contains about 204,200 cubic metres of process and other wastes and about 147,000 cubic metres of slightly contaminated soil. The facility is worrisomely close to the eroding cliffs of the lake.

Contamination was spread to locations within Port Hope in a number of ways, including deposition in undesignated sites, use of contaminated materials as fill, spillage from haul vehicles, and wind and water transport from storage sites. The problem of radioactive contamination in the Town was recognized in the mid-1970s and since that time a number of programs have been undertaken to identify contamination, clean up contaminated areas, and consolidate waste into controlled areas. Some contamination also lies at the bottom of the harbour. The amount of waste in Port Hope is about 173,000 cubic metres of waste and 92,000 cubic metres of slightly contaminated soils.

Other Historical Wastes in Canada

About 8,000 cubic metre of marginally contaminated soils are stored in the Toronto suburb of Scarborough as a result of a cleanup conducted in 1996. These wastes arose from a private radium dial-painting facility that operated at the time of the World War II.

An industrial site in Surrey, British Columbia, was discovered to have a small amount of historic radioactive wastes (5,000 cubic metres) consisting of thorium-contaminated slag from the smelting of imported niobium ore during the 1970s. A clean-up was conducted and most of the contaminated soils were sent for disposal at a site in Oregon with a smaller amount sent to AECL's Chalk River facility.

In 1992, uranium-contaminated soil and building material were discovered in an unused warehouse in Fort McMurray, Alberta. This was the southern terminus of a long water transportation route from the fort Port Radium uranium mine in the Northwest Territories that was used in the 1930 to 1950s. Remediation has been completed in Fort McMurray, but in 2008 work was still being done along the transportation corridor.

The Low-Level Radioactive Waste Management Office

This office was formed in 1982 and is managed by Atomic Energy of Canada on behalf of Natural Resources Canada who provide funding and set policy. The mandate is to deal with the historic radioactive wastes described above and to provide leadership for managing all other low-level radioactive waste in Canada. The largest project is the work to remediate the Port Hope, Ontario. Smaller clean-up projects have been conducted in Scarborough, Ontario; Fort McMurray, Alberta; Surrey, British Columbia; and the transportation route from Port Radium, Northwest Territories, to Fort McMurray, Alberta.

In the early 1980s, efforts were made to find a permanent solution for the historic wastes at Port Hope. Negotiations to find a disposal site broke down, however, and the federal government formed a Siting Process Task Force in 1986. Its mandate was to develop a process for siting a disposal facility in Ontario for the Port Hope wastes and possibly other wastes. In 1988, the Task Force recommended that siting be based on communities volunteering sites rather than the government dictating locations. In addition, volunteer communities would have considerable involvement in the development of the facility and would receive compensation.

In early 1989, a Siting Task Force was formed to implement the recommendations. In 1996, following extensive discussions with numerous communities in Ontario, only one—Deep River, near the Chalk River Laboratories—agreed to host the disposal site. The disposal facility was to be at the Chalk River Laboratories site and would consist of a mined cavern. However, negotiations between Deep River and the federal government broke down.

With no place in Ontario willing to host a disposal facility, the communities in the Port Hope area felt that the wastes would not be properly dealt with unless they hosted the facility. An agreement was reached, and in 2001 the federal government announced it would allocate $260 million to construct new disposal facilities under a program called the Port Hope Area Initiative. The initiative consists of two projects. The first is called the Port Hope project and it involves constructing a new engineered surface mound facility at the Welcome site. The wastes currently at the Welcome site as well as all the wastes in Port Hope would be placed in the mound, which would be sealed and monitored. In 2008, the project was in the licensing phase with a target of completing the project by about 2020.

The Port Granby project is similar and will relocate the wastes currently stored adjacent to a new engineered surface mound that is further from the eroding shoreline cliffs. It is expected the project will be completed by about 2016.

The Swedish Approach to Low-Level Radioactive Waste Disposal

As Canada has not yet implemented final disposal of low-level waste, it is instructive to look at what is being done in other countries. Sweden has developed the most unique and advanced disposal facility in the world, for either nuclear or non-nuclear waste. Because the geologies of Canada and Sweden are similar, this approach could be adapted for use in Canada.

The Swedish Final Repository for low-level waste is located on the east coast at the Forsmark Nuclear Power Station about 160 kilometres north of Stockholm. What makes the disposal method unique is, first, that it is located underground and, second, that it is situated under the Baltic Sea. Access to the repository is via two tunnels whose entrance is on land and which slope gently downward for a distance of one kilometre underneath the Baltic Sea. One of the tunnels is dedicated to waste transportation and is equipped with remotely-controlled vehicles. The repository consists of a large vertical silo (60 metres high and 30 metres in diameter) and four parallel, horizontal caverns. The top of the repository is about 50 metres below the sea bed and the depth of the Baltic Sea in this area is about 5 metres (Figure 9-3).

The repository started to accept wastes in April 1988. The facility will take all of the low-level radioactive waste generated in Sweden to approximately the year 2010, approximately 90,000 cubic metres. This is less than five per cent of the capacity of a single medium-sized municipal landfill (of which there are thousands in Canada), and illustrates the "small" size of nuclear wastes.

The Swedish Final Repository is like a box within a box within a box within a box. The first barrier is the waste form itself. Before transportation to the repository, the waste is mixed with cement or asphalt and placed in concrete or metal containers.

The silo, which has 0.8-metre-thick reinforced-concrete walls, contains the most highly radioactive materials. The space between the concrete silo and the rock wall is filled with bentonite clay, which expands when it becomes wet. As each layer of waste is emplaced, it is grouted permanently into place with concrete. Thus the engineered barriers consist of the immobilized waste form, grout, the silo wall, and the bentonite layer. Similar barrier systems are used in the four horizontal caverns. The surrounding rock mass provides an additional barrier.

The wastes inside the Swedish Final Repository will decay to about the same level of radioactivity as the surrounding rocks in about 500 years. Less than 10 percent of the radioactivity will remain after one hundred years. This is about the same length of time as it takes for organic wastes to decompose in a non-nuclear municipal or industrial landfill.

The Swedish repository utilizes only minimal land surface and places no burden on future generations.

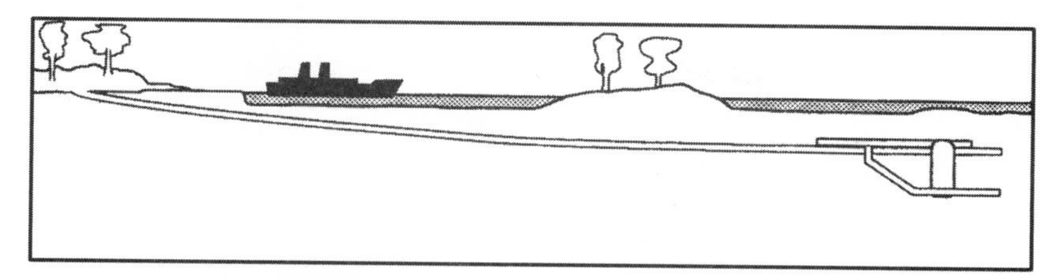

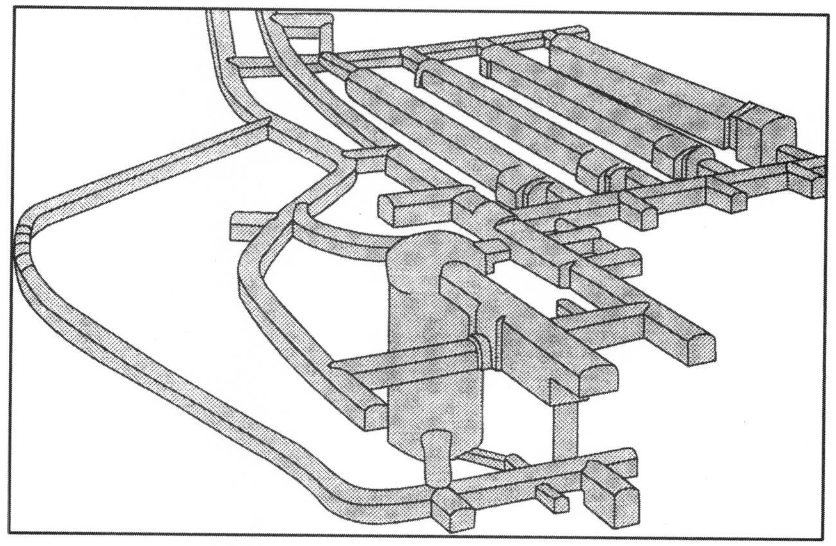

FIGURE 9-3: The Swedish low-level radioactive waste facility at Forsmark Sweden. As shown in the top panel, entry to the depository is by tunnels that extend under the Baltic Sea. The waste is deposited in the large silo and in the four horizontal caverns shown in the lower panel.

In Finland, underground caverns are used for disposal of low-level waste at their two reactor sites; the geology consists of precambrian hard similar to that at the Swedish disposal facility. The United States has been operating an underground waste disposal facility in a salt formation in Nevada since 1999.

In summary, although more heterogeneous than high-level wastes, low-level waste are relatively small in quantity, which allows innovative treatment and disposal methods to be used.

-10-

High-Level Nuclear Waste

With their valves, refuelling machines, coolant pumps, computer controls, temperature and neutron-flux monitors, and a host of other precision-made components, it is clear that nuclear reactors are complex machines. It is hard to imagine that nuclear fission could ever occur spontaneously in nature. But this is exactly what happened long, long before humans unlocked the atom.

In 1972, French scientists discovered that at least 17 "natural" nuclear reactors had operated in the rich Oklo uranium deposit in Gabon in West Africa. This was possible because, as we have seen, radioactive materials decay over time naturally; at an earlier time in Earth's history, the planet contained much more radioactivity. In particular, 1.8 billion years ago, the concentration of uranium-235 was about 3.7 percent (its present concentration is 0.7 percent; it has a half-life of 710 million years), which is similar to the concentration used in today's light-water reactors. Groundwater flowing through the deposit acted as a neutron moderator; uranium fission reactions started spontaneously and continued for hundreds of thousands of years. Scientists have determined that each of the "reactors" operated at about 20 kilowatts thermal power.

The Oklo site has been extensively studied because it provides valuable information about how radioactive products of these early reactors—that is, spent fuel or high-level radioactive waste—interact with the surrounding environment. The radioactivity that was created (a total of about 5.4 tonnes of fission products and 1.5 tonnes of plutonium) have long ago decayed to stable elements. Studies indicate that these stable daughter products have not migrated from the site, and have remained remarkably immobile—a surprising result given the abundant groundwater and the shallow depth of the ore body, not to mention that the "wastes" were not immobilized into a solid, leach-resistant form, as is proposed for today's nuclear waste. It appears that the clays and bitumen present at the site played an important role in containing the waste. Nature has demonstrated that nuclear wastes can be safely contained without harming the environment.

Nuclear Fuel Wastes

In Canada, high-level radioactive waste consists of spent fuel, which has been irradiated in a reactor until about 67 percent of the uranium-235 is used up and power can no longer be produced effectively. This takes about one and a half years in CANDU reactors. The spent fuel is sometimes referred to as used fuel or nuclear fuel waste.

Spent fuel has three important characteristics. First, it is highly radioactive. Second, it is very small in volume compared to wastes created by many other industries or by burning coal for energy. Third, the waste is contained; it is not emitted into the environment.

Spent fuel contains two types of radioactive nuclides that have been created during the time in the reactor: fission products and actinides. Fission products result from uranium-235 atoms splitting when hit by neutrons. Since the uranium atom does not always split in exactly the same way, several dozen different isotopes of approximately half the atomic weight of uranium are formed. These have relatively short half-lives, ranging from seconds to several tens of years. Some of the more prominent fission products are strontium-90, cesium-137, and krypton-85. Fission products initially generate large amounts of radiation and heat, so fuel bundles must be handled remotely; they must also be shielded and cooled when removed from the reactor.

The second type consists of nuclides that absorb one or more neutrons, but do not fission. Instead, they transform into nuclides which have slightly greater atomic weight than uranium, called actinides (or transuranics) such as plutonium-239, americium-241, and neptunium-237. In general the actinides have very long half-lives and low activity. Plutonium-239, for example, has a half-life of about 24,100 years.

The composition of a typical CANDU fuel bundle before and after being in the reactor is shown in Table 10-1. The major change is the transformation of about two-thirds of the uranium-235 to fission products. In addition, there is an intermediate reaction in which a small amount (less than 1 percent) of the uranium-238 absorbs a neutron and transforms to plutonium-239, of which just over half fissions to produce additional fission products. About 50 percent of the energy derived from a CANDU fuel bundle is derived from these plutonium fissions.

Table 10-1 CANDU Fuel Bundle Composition

Isotope	Fresh	Removed from Reactor
uranium-238	99.30%	98.70%
uranium-235	0.70%	0.23%
fission products	–	0.80%
plutonium-239	–	0.27%

Given the enormous energy created by the fuel bundle, it is surprising how little of the material inside the fuel actually changes. The spent bundle looks identical to the original bundle, and only about 1.1 percent of the material inside has been modified. This is dramatically different from coal, oil, and other fuels that undergo a complete physical transformation during the combustion process.

It should be noted that the fissile nuclides, plutonium-239 and uranium-235, together form about 0.5 percent of the spent fuel. This is only a fraction less than the original fissile content (0.7 percent of uranium-235). For this reason, the spent fuel is not necessarily a waste, as these fissile materials could be extracted by reprocessing and used to generate more power. The various fuel cycles in which this fissionable material could be used are discussed in Chapter 14. Liquid wastes generated by such reprocessing are high-level wastes. They could be solidified into glass or ceramic blocks prior to long-term storage or disposal. Because the extracted plutonium could be diverted to making nuclear bombs, reprocessing is not being performed in Canada, nor are there plans for this in the future.

Interim Storage

A 720-megawatt CANDU reactor produces about 100 tonnes (about 4,200 bundles) of spent fuel per year. The fuel bundles are removed from the reactor core by remote control and placed into deep water-

FIGURE 10-1: Pool storage bay for spent fuel. Spent-fuel bundles are placed into baskets and racks with about four metres of water above the top of the fuel racks. The water, which emits a distinctive blue glow from Cerenkov radiation, shields station personnel against radiation.

filled pools located at the reactor sites (Figure 10-1), which provide cooling as well as shielding. All the spent fuel produced in Canada until 2008 would fill five hockey rinks to the height of the boards.

After the bundles have been in the pool for about seven years they can be transferred to dry storage. In Canada, large concrete structures with 1-metre-thick walls have been developed that can remain outdoors. These structures are being used to store spent fuel from the closed Douglas Point reactor as well as from the Point Lepreau and Gentilly-2 generating stations (see Figure 10-2). There are two buildings at the Western Waste Management Facility at the Bruce Nuclear Power Development site, each with a capacity to contain 500 concrete casks; each of these casks holds 384 fuel bundles. In the United States, steel as

FIGURE 10-2: The dry used-fuel storage facility at Hydro Quebec's Gentilly-2 nuclear station. The concrete structures can safely store fuel for decades. The overhead crane system is used for remote handling of the radioactive bundles.

well as concrete canisters are being developed. The water pool and dry storage canisters can store spent fuel for periods exceeding 50 years.

One option is to move used fuel from reactor swimming pools to a central location for long-term storage, possibly up to 50 years or longer, prior to placing the fuel into a final disposal repository.

In the United States, such a facility—called the monitored retrievable storage facility—was a key part of their fuel cycle strategy. This was because many of their reactors were going to run out of on-site pool storage capacity before a permanent disposal facility would be ready. Although the monitored retrievable storage facility was scheduled to begin accepting spent fuel in 1998, the issue of where to site the facility met with public opposition, and the project was cancelled.

In Sweden, an underground central spent-fuel storage facility came into operation in 1985. The facility, known as "CLAB," will ensure all spent fuel from the Swedish nuclear power program can be safely stored until a permanent disposal facility is ready. The facility uses storage pools much like those at a reactor site, except that they are located underground in the granitic rocks of the Baltic Shield.

Permanent Disposal

Those countries operating nuclear power reactors have studied how high-level radioactive wastes should be disposed and agree that they should be enclosed in a solid leach-resistant form and then buried deep in a stable geologic formation. At present, no country has yet constructed a repository for high-level nuclear wastes (although three geologic repositories are in operation for low-level wastes; see Chapter 9). A variety of different geologies are being investigated. Belgium, for example, is studying clay formations. The United States is investigating the suitability of the volcanic tuffs of Yucca Mountain in Nevada.

Finland is the closest to implementing disposal of high-level nuclear wastes. In 2001, the Finnish parliament approved the plan to build a repository in the crystalline Precambrian rocks of southwest Finland near the nuclear power plant at Olkiluoto. Construction began in 2004 and if the exploratory phase proves positive, the site will be expanded into a repository at a depth of approximately 500m, accessed by a shaft and spiralling tunnel. If licensing approvals are obtained, it could begin to receive waste by about 2020.

At this time there is no urgency to build a permanent disposal facility in Canada because dry storage facilities can provide safe storage for many decades. Nevertheless, the federal government deemed it necessary to demonstrate the feasibility of permanent disposal. Therefore, a research program was launched in 1975 to develop a concept for disposing of high-level nuclear wastes deep underground. Led by Atomic Energy of Canada Limited (AECL), it was known as the Canadian Nuclear Fuel Waste Disposal Program and conducted research until about 1998.

A broad range of disposal options has been considered by AECL and the international community, including exotic solutions such as shooting nuclear wastes into space and burying them in polar ice caps or below the seabed. Although some of these solutions are theoretically possible, there are many practical challenges. Shooting wastes into space, for example, is prohibitively expensive and also poses risks, as witnessed by the tragic explosion of the space shuttle Challenger in January 1986. Furthermore, international treaties prohibit Canada from disposing of wastes in international territories including the oceans and Antarctica.

Given Canada's geography and geology, the stable ancient rocks of the Canadian Shield were selected as the most suitable host, a decision supported by three independent review groups. The keys to safety are placing the wastes deep underground and in a solid, leach-resistant form. These simple yet crucial steps remove the wastes from the forces of erosion and caprices of human intrusion; it also preserves surface land space for other beneficial uses for the growing population.

The repository in AECL's disposal concept consists of a grid of tunnels at a depth of about 500 to 1,000 metres. Waste containers would be placed in vertical boreholes drilled into the floors of the tunnels. The boreholes would be plugged with materials such as bentonite clay. Once the waste is in place, the tunnels, shafts, and exploratory boreholes would be sealed and grouted. The conceptual repository is shown in Figure 10-3. An alternate design was also developed, in which waste containers are placed horizontally in the tunnels and surrounded by bentonite clay.

A key aspect is that the waste would be put into a very inert, stable, solid form before being sealed underground. Research indicated that titanium or copper containers would have the necessary corrosion-resistant properties. Glass beads would be compacted into the spaces between the fuel and the container shell to provide internal support against underground pressures. These containers, which are cylindrical in shape and contain 72 fuel bundles, have been designed so no waste would be released from them for at least 500 years.

Research was also conducted into the materials and methods for sealing the repository and associated shafts and boreholes. A mixture of clay and sand packed around the waste containers would limit groundwater access and would trap radioactive material that might escape. On contact with water, the clay would swell and provide a seal around the containers.

The main barriers to prevent release of radioactivity would consist of:

- the waste form, i.e. ceramic uranium dioxide fuel pellets
- the waste container
- buffer material surrounding the container
- the geologic formation.

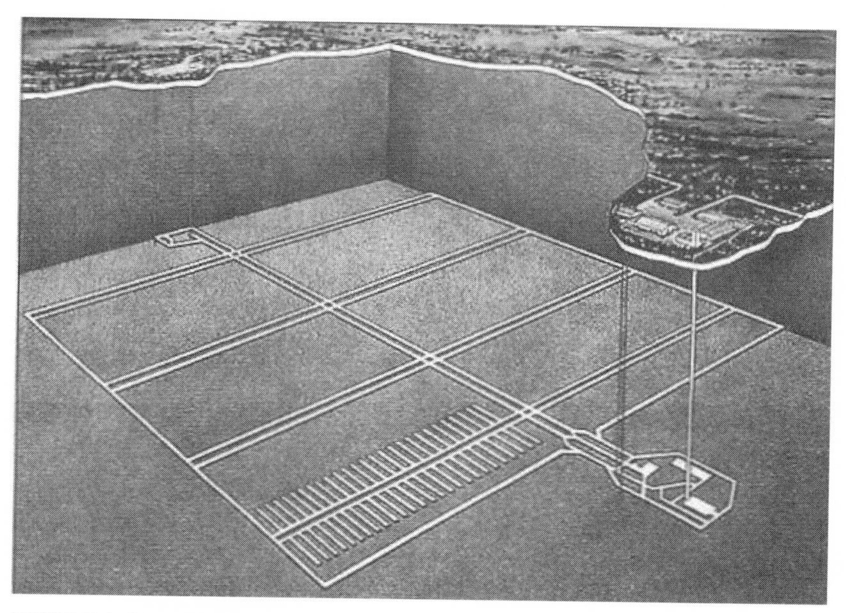

FIGURE 10-3: Conceptual design of a spent-fuel disposal facility. Fuel bundles sealed in appropriate containers would be deposited in disposal rooms, which would then be backfilled and sealed.

The only possible method by which the nuclear wastes could become harmful to humans and the environment would be if they were to leach out of the canisters, and then be carried by groundwater to the earth's surface in concentrations that are sufficient to cause harm.

But this is virtually impossible. The main reason is the special character of the Canadian Shield. These rocks are amongst the oldest and most stable in the world. Unlike geologic areas with limestone that is soluble, cavities do not exist in Canadian Shield granite. In hundreds of kilometres of drilling in granitic rocks of the Shield, only cracks (no cavities) have been observed, of which the largest are only a few millimetres in width. This is in spite of the very long time that ground water has been present in these rocks (a large portion of the Canadian Shield is roughly 2,500 million years in age).

For nuclear waste disposal, the period of concern is 10,000 years, a small length of time compared to the age of the Shield. Even if the natural leaching process were increased a thousand-fold by the heat generated by the waste, simple calculations show that only trivially small openings would be leached over 10,000 years. The reasons are that, first, granitic rock is composed of essentially insoluble silicate minerals, similar to glass. Second, although fractures exist, water resides mostly at the surface because the permeability and porosity of these cracks decrease with depth as the pressure of the overlying rock mass increases. Little groundwater exists at the depth of the proposed disposal vaults; any water which is present is very old and very slow moving.

As long as the job is done carefully and thoroughly, granitic formations should safely entomb nuclear wastes for millions of years. A suitable geologic formation must be selected, wastes must be immobilized in solid leach-resistant form, and the tunnels and shafts of the repository must be properly sealed once full.

The main way of demonstrating the long-term safety of this disposal concept was through sophisticated computer programs that model the movement of radionuclides as they are leached from the waste containers and carried by groundwater from the repository through the rock mass to food chains. These analyses, known as a risk assessment, are complex, as all the pathways by which radionuclides may reach humans must be identified and modelled.

Much of the data for the computer analyses was obtained from the Underground Research Laboratory located in the Lac du Bonnet batholith, a large granite rock mass near Atomic Energy of Canada Limited's Whiteshell Laboratories in Manitoba. A shaft was sunk to a depth of 445 metres in previously unmined rock, and a number of galleries and rooms were excavated in which various experiments were conducted by AECL. Starting in 1982, experiments studied the effects of heat on rock, the transport of radionuclides through rock fractures, methods for sealing rooms and boreholes, methods of mining and blasting that minimize damage to the surrounding rock, and much more. The Laboratory ceased operation in 2005.

The risk assessments provided quantitative predictions of potential radiation doses to humans. The results showed that the concept is safe and would meet all regulatory requirements.

Considerable national and international review has gone into the Canadian program. For example, Japan, the United States, Sweden, and France joined in cooperative research programs at the Underground Research Laboratory. The Technical Advisory Committee was formed in 1980 to provide independent review of the program. This Committee was composed of distinguished scientists and engineers nominated by various Canadian scientific organizations and had no direct links with AECL or the nuclear utilities.

Regulation

The Canadian Nuclear Safety Commission is the regulatory agency responsible for licensing all nuclear matters in Canada, including the disposal of spent fuel. It has issued a series of regulatory guides and monitors the national high-level waste disposal program. The CNSC requires that the burden on future generations be minimized by:

- selecting disposal options that do not rely on long-term institutional controls;
- ensuring that there are no predicted future risks to human health and the environment that would not be currently accepted;
- ensuring that the radiological risk does not exceed one in a million serious health effects per year.

The regulations imposed on nuclear wastes are far more stringent than those for non-nuclear wastes. A main difference is that computer-based risk assessment is mandatory for licensing a nuclear waste disposal facility but is not required for licensing hazardous-waste disposal facilities. Another difference lies in the consideration of future generations. Nuclear waste regulations require that disposal methods must not rely on long-term institutional controls; mathematical risk assessment is to be performed for up to 10,000 years into the future.

In contrast, municipal landfills in Ontario, for example, seldom predict impacts more than a few decades into the future. Post-closure guidelines stipulate the primary liner should have a lifetime of 1,000 years. The Ontario Ministry of Environment's approval is required if a landfill site is to be used for other purposes within 25 years of its closure. There is also a requirement to maintain engineered works such as leachate collection systems "as long as needed." No discussion is provided as what might constitute "as long as needed." Financial assurance must be established and monitoring be conducted for "as long as contaminants from the site pose a potential concern to the environment." In a few recent cases, a financial assurance period of slightly over a century has been stipulated by the Ministry.

The situation is the same for hazardous wastes. The Canadian Council of Ministers of the Environment guidance document makes little mention of the post-closure period other than to state that "hazardous wastes can retain their harmful properties over a long period of time, perhaps for centuries." The Council recommends that a post-closure program be implemented that includes insurance, maintenance, contingency plans, and monitoring. Depending on site conditions, it states that this post-closure period "could well exceed 100 years."

Unlike nuclear waste disposal, no systematic thinking has gone into the long-term implications of municipal and hazardous landfills. There are no definitions of the post-closure period, nor has there been consideration of long-term issues such as ice-ages, seismic activity, or future human intrusion, which are routinely assessed for nuclear wastes.

Furthermore, because the facilities are located at the surface, perpetual on-going maintenance will be required. Put bluntly, municipal, industrial, and hazardous landfills only address the short term and are designed and regulated to give the problem to future generations.

The Canadian Environmental Assessment Agency Panel Review

AECL's disposal concept for nuclear fuel waste was referred for review under the federal environmental assessment review process in 1988. Following numerous public hearings that included 561 written submissions, the Panel finally presented its results in 1998. This review broke new ground in two ways: it was the first time a concept, rather than an actual, site-specific project was subjected to a federal environmental assessment, and second, it lasted far longer (ten years) than any other environmental assessment. The federal government clearly showed that it did not want to deal with this politically sensitive issue.

The panel concluded that the safety of the concept had been adequately demonstrated from a technical point of view, but not from a social perspective. As the concept did not have broad public support, it therefore was not acceptable.

A number of recommendations were presented, including the need for:

- a First Nations participation process;
- a comprehensive public participation process;
- an ethical and social assessment framework;
- an arms-length agency to manage nuclear waste disposal, separate from AECL and the nuclear utilities.

The panel recommended that the search for a specific disposal site should not be pursued until these recommendations are implemented and broad public acceptance of the nuclear waste disposal concept is achieved.

In response, the federal government introduced the Nuclear Fuel Waste Act in April 2001, which required the nuclear utilities to establish a separate organization and a dedicated fund for the long-term management and disposal of nuclear fuel wastes. The new waste management agency was to assess disposal alternatives and establish a comprehensive public participation program.

As a result, the Nuclear Waste Management Office was established in 2002 by Ontario Power Generation, Hydro Quebec, and New Brunswick Power. It embarked on a five-year study that assessed different spent-fuel disposal options including deep geologic disposal, continued storage at reactor sites, and centralized storage.

In 2007, the government accepted the recommendation from this Waste Management office to proceed with "adaptive phased management," a process of selecting a centralized site (which could take 30 years), and developing a storage/disposal site (which might be above or below ground). Spent fuel would be stored and monitored in a retrievable manner until society feels it is time for disposal; this could be up to 300 years or more. The facility might take the form of AECL's conceptual deep repository but without the sealing and backfilling, so that the spent fuel remains monitorable and retrievable; it might be a temporary shallow cavern such as Sweden's CLAB; or it might be a surface facility. Thorough public consultation and involvement at every step is a cornerstone of the project.

It is the longest project ever initiated in Canada, one that will drag on for many decades, possibly even a century or more.

The public perceives nuclear wastes to be uniquely toxic and extremely long-lived; there is also the perception that scientists are unable to find methods for their disposal. A comparison shows that many non-nuclear wastes have similar toxicity, are even more long lived (if not infinite), and are much greater in volume. Furthermore, Canada's non-nuclear hazardous wastes are disposed of in surface landfills without encapsulation. Recently, proposals have arisen for massive injection of carbon dioxide in gaseous form into deep geologic formations. In comparison, the methods proposed for nuclear waste disposal are far superior to the way in which we dispose of hazardous waste (or the proposed disposal of massive quantities of carbon dioxide).

-11-

Nuclear Medicine

The practice of medicine has been profoundly transformed by nuclear technology. Where surgery was previously necessary for doctors to peer inside the body to diagnose and treat problems, now a host of nuclear methods are used. Today, the technology is so prevalent that one out of three Canadians has undergone a nuclear procedure of some kind. A new medical branch, nuclear medicine, has evolved over the past four decades, and Canada has been at the forefront in developing and using the high-technology nuclear tools that made it possible. In 1969, for example, Quebec doctors were the first in the world to recognize nuclear medicine as a separate specialty.

Diagnostic Nuclear Medicine

Nuclear medicine is composed of two distinct parts: diagnosis (understanding the nature of the disease) and therapy (treatment). The large majority of nuclear medicine procedures are diagnostic. For example, coronary artery disease is the most common type of heart disease, and is the leading cause of death in North America. Radiopharmaceuticals and gamma cameras are among the hundreds of diagnostic nuclear procedures used by cardiologists to diagnose this disease.

X-rays

The discovery of X-rays has had an enormous impact, becoming the primary diagnostic tool of modern medical and dental practice. Many of us have had our teeth X-rayed, and X-rays of fractured bones are routine. The shadowy images that appear on the photographic film reveal dental cavities or fractures against the denser bone. Subtle differences in other organs can also be detected and provide clues about the presence of disease.

X-rays are photons resulting from the collisions of electrons with atoms. Strictly speaking, X-rays are not a nuclear but an atomic

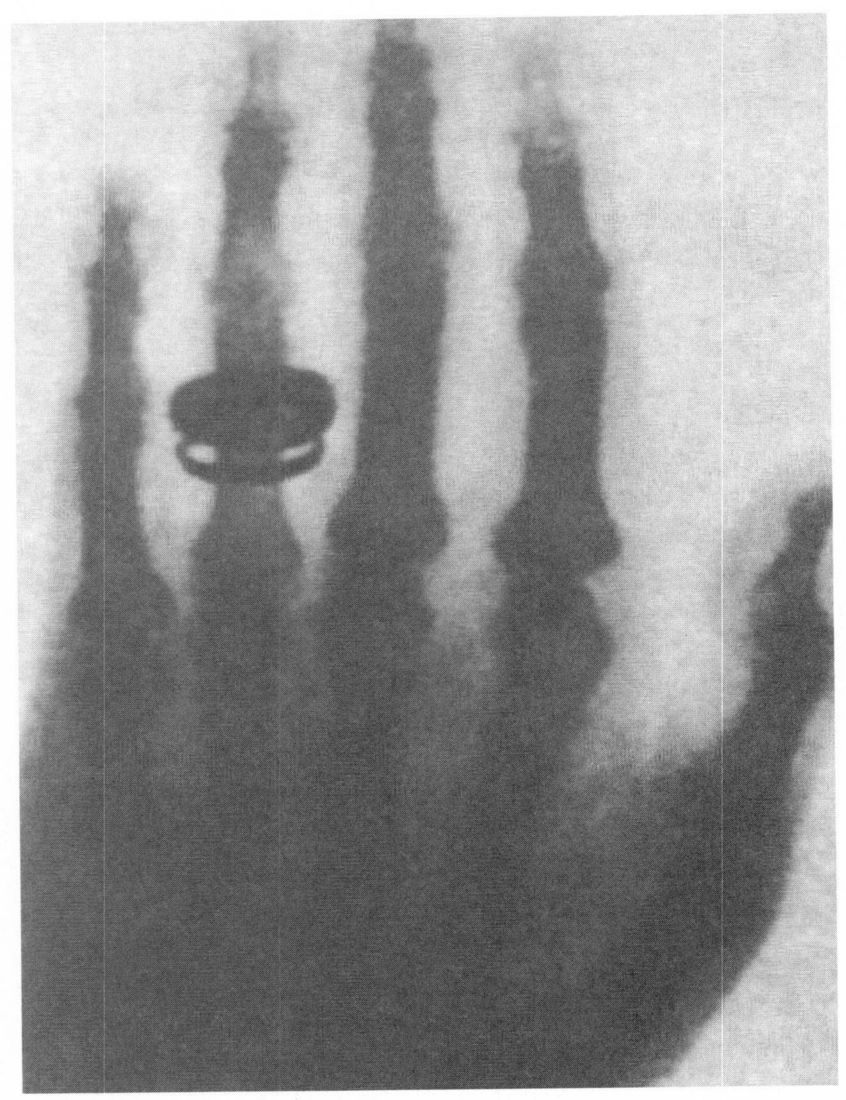

FIGURE 11-1: The first X-ray picture ever published. It was taken by Wilhelm Roentgen of Germany in 1896 and shows his wife's hand. For this discovery Roentgen received the first Nobel Prize in Physics in 1901.

phenomenon (although they are physically the same as gamma rays). The difference is that X-rays originate from changes in the energy levels of atomic electrons and gamma rays from the energy levels of nuclei.

X-rays are one of the most frequently used tools in medical practice. For example, in Ontario alone about 41,000 X-rays are taken each day

(Note 11-1). Dental X-rays are also common. Radiation doses range from approximately 0.02 mSv for a dental X-ray to about 0.07 mSv for a chest X-ray. X-rays are also emitted by older television screens and computer monitors, although the doses to individuals are small.

Radioactive Tracers

The use of radioisotopes in diagnostic medicine began in the late 1940s with the use of radioactive iodine to study the function of the thyroid gland and to diagnose thyroid disease. With experience, the use of radioactive elements to study the metabolic processes in the body led to breakthroughs in diagnostic medicine. In the 1960s, the specialty discipline of nuclear medicine grew rapidly. Initially, techniques were developed to measure blood flow to the lungs and to identify cancer "hot spots." By the 1970s, most of the body's organs could be studied, including the liver and spleen, the brain, and the gastrointestinal tract. In the 1980s, radiopharmaceuticals were designed for diagnoses of heart disease and cancer. Many further advances have been made, some of which are described below.

Diagnostic nuclear medicine is possible because the nuclear decay process emits particles that can be detected from a distance. An appropriately radioactive-tagged drug, called a "radiopharmaceutical," is introduced into the body either by injection or orally. The drug is selected to be absorbed by the specific organ under examination. Today some 20 to 50 radiopharmaceuticals are in routine use for diagnosis, most of which are organic in nature. The path of the radiopharmaceutical is followed by scanning the relevant organ with a special instrument called a "gamma camera." This is able to detect even minute amounts of gamma radiation, a procedure that avoids the previous practice of potentially traumatic exploratory surgery. Nuclear medicine is rendering the scalpel obsolete as a tool for diagnosis.

It should be noted that radioactive tracers provide information on how organs function. In this way they complement X-rays, which show the shape of organs.

Although there are more than 1,500 known radionuclides, only a few of these are suitable for medical diagnosis. The primary requirement is to minimize the patient's exposure to radiation dose by using gamma rays with just enough energy to be detectable outside the patient's body and with radiation that lasts just long enough—that is with a short half-life—to allow completion of the diagnostic tests. These radionuclides are mixed with chemical compounds that are attracted to the organs of interest. Radioisotopes such as technetium-99m, iodine-131, and thallium-201 are used to study the heart, brain, bones, lung, thyroid, liver, and kidneys (see Table 11-1). Approximately 40,000 procedures are performed each day in North America. Canada supplies about 50 percent of the world's medical isotopes.

Table 11-1 Isotopes Used in Medical Diagnosis

Isotope	Organ System(s)
Technetium-99m	brain, lung, bones, heart, infection, tumour
Thallium-201	heart
Iodine-131	thyroid
Iodine-125	blood, radioimmunoassay, prostate
Iodine-123	thyroid, heart, brain
Xenon-133	lungs
Gallium-67	soft tissues
Indium-111	infection, inflammation
Strontium-82*	heart

* produces rubidium-82

In nuclear medicine, the most widely used isotope is technetium-99m (Tc-99m). It has a low-energy gamma and short half-life (six hours), so it does not stay in the body very long. Discovered in 1937, it came into general usage in 1973 when Atomic Energy of Canada Limited's Chalk River Laboratories began regular production of its parent, molybdenum-99.

A chemically flexible isotope, Tc-99m can be formulated into numerous compounds for a variety of functions. For example, when combined with albumin particles and injected intravenously, it is trapped in the blood vessels of the lungs, and helps to identify areas with decreased or absent blood flow. Linked to a phosphate molecule, it will concentrate in bone tissue at a rate proportional to blood flow and bone production level. More than twenty different technetium compounds are used, and it is estimated that every year, some 30 million people worldwide undergo Tc-99m diagnostic procedures (Note 11-2).

The Travels of Moly

Molybdenum is the parent of technetium, the most important and widely used nuclear tracer in medicine. Its delivery to hospitals around the world is a formidable exercise in logistics, since molybdenum has a half-life of only 66 hours.

Molybdenum is produced by irradiating a uranium-aluminum alloy in which the uranium is highly enriched for about two weeks in the NRU reactor at Chalk River Laboratories. Molybdenum-99 is created as one of the fission products. The irradiated uranium-alloy piecess are removed from the reactor in the afternoon and are processed by remote control in heavily shielded hot cells. The pieces are dissolved and the resulting solution is passed through an ion-exchange column to extract the molybdenum-99.

The molybdenum-99 is rushed to Ottawa, about a two-hour drive, where MDS Nordion technicians work overnight to do the final processing and packaging. Early the next morning, the molybdenum processing and testing are complete and the product is packaged for shipment. The molybdenum is then delivered to the Ottawa airport and flown by chartered flight to the US radiopharmaceutical producer that uses the molybdenum-99 to make technetium-99m generators. Shipments to Europe, Japan, and South America are made by scheduled cargo flights. Hospitals are supplied with portable lead-lined "generators" from which technetium-99m is drawn off, or "milked," as needed.

Canada is the world's leading supplier of molybdenum-99, providing over 50 percent of the world's supply.

Because of its very short half-life, technetium itself cannot be shipped to hospitals. Instead, its parent molybdenum-99 is manufactured and shipped across the world. Figure 11-2 shows how molybdenum is packaged for shipping. Most of the world's supply is specially made in the NRU reactor. The sidebar describes the highly coordinated process

FIGURE 11-2: Container system for shipping radioactive molybdenum. Note that several containers are layered one within the other for maximum safety. About a million shipments of radioisotopes are made each year in Canada using such containers.

of delivering this important radioisotope to hospitals around the world in a timely fashion.

The Gamma Camera

The workhorse of most nuclear medicine departments is the gamma, or Anger, camera. The camera is composed of four main components: a collimator, a scintillation crystal, a light detection system, and a computer processor. Figure 11-3 shows how gamma rays are emitted from an organ that has been "tagged" with an appropriate radionuclide. The collimator is made of lead and contains a series of parallel holes that ensure that only those gamma rays travelling parallel to the holes pass through. The crystal is composed of sodium iodide which contains thallium. This crystal has the special property that when a gamma ray passes through, it scintillates—or flashes—a blue light whose intensity is proportional to the energy of the gamma ray. Behind the crystal is a layer of photomultiplier tubes that convert the light flash to a measurable electric voltage; this is proportional to the size of the light flash and therefore to the energy of the gamma ray. Finally, the data is processed by computer to produce an image of the activity in the organ. By positioning the camera in the right place, the source of the gamma rays can be determined. Computer techniques are used to form an image showing where the radioisotope is located. The image can then be used to deduce how that organ is functioning.

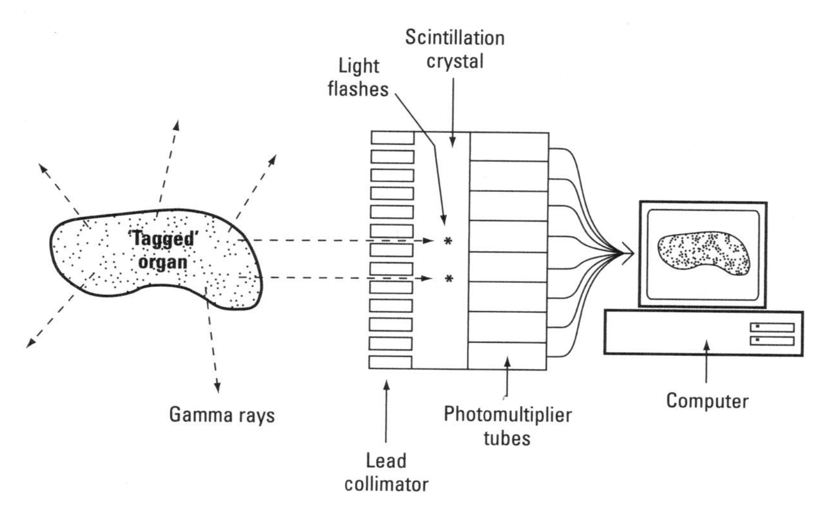

FIGURE 11-3: Schematic of a gamma camera. Gamma rays from an organ tagged with a radiochemical cause light flashes in a sodium iodide crystal that are detected by photomultipliers that send the electronic pulses to a computer for analysis and display.

A drawback of the gamma camera is that it only gives a two-dimensional picture. Two other methods, Positron Emission Tomography (PET) and Single Photon Emission Computerized Tomography (SPECT), have been developed that allow three-dimensional views.

Nuclear diagnosis techniques are not particularly useful in demonstrating anatomy, but are very powerful in demonstrating the function of different organs.

In Vitro Methods

It is very common for a patient's blood or tissue to be analyzed in a laboratory, with no radiation dose to the patient. This is known as *in vitro* (Latin for "in glass," in other words using a test tube), compared to *in vivo* (in a living person) techniques. A method called radioimmunoassay uses radioisotopes to detect evidence of cancer by revealing the presence of antigens, or tumour markers; these can indicate the presence of cancerous cells. This method is infinitely more accurate than any previous technique. A radioactively labelled substance (an antibody, for example) is mixed with a sample of the patient's blood; this reacts with the compound of interest. Increasing concentrations of non-radioactive substance are added and the amount of labelled substance displaced from the compound is measured. This allows the concentration of the compound of interest to be calculated very accurately. In 1977, Dr. Rosalyn Yalow won the Nobel Prize in medicine for developing the radioimmunoassay method and applying it to insulin. Radioimmunoassay can also be used to detect infectious diseases and measure minute quantities of hormones, vitamins, and drugs in bodily fluids.

Positron Emission Tomography (PET)

Positron Emission Tomography is used to study subtle chemical changes and metabolism in the brain and other organs. A radiochemical that emits positrons is taken up by the organ under examination. A positron is very unstable. It is immediately annihilated by interacting with an electron; this creates two gamma rays that travel in opposite directions. A ring of gamma-ray detectors is placed around the part of the body, close to the organ under study. A pair of detectors on opposite sides of the ring each identify a gamma ray at the same instant; a positron decay must have occurred on the line joining the two detectors. A ring of detectors is used that is capable of detecting many annihilation events. If a sufficient number of lines of interaction are identified, then an image of the distribution of radioactivity can be constructed.

As an example, PET scans can view the distribution of fluorine-18 in a two-dimensional plane that transects the head (see Figure 11-4). By measuring several planes, a three-dimensional picture can be constructed. The extent to which different regions of the brain are functioning and the response of brain tissue to external stimuli can be tested in both normal and diseased states. For example, an epileptic fit suffered by a child shows up as a dramatic increase in activity in a given region of the brain. For this reason, PET scans are used to study

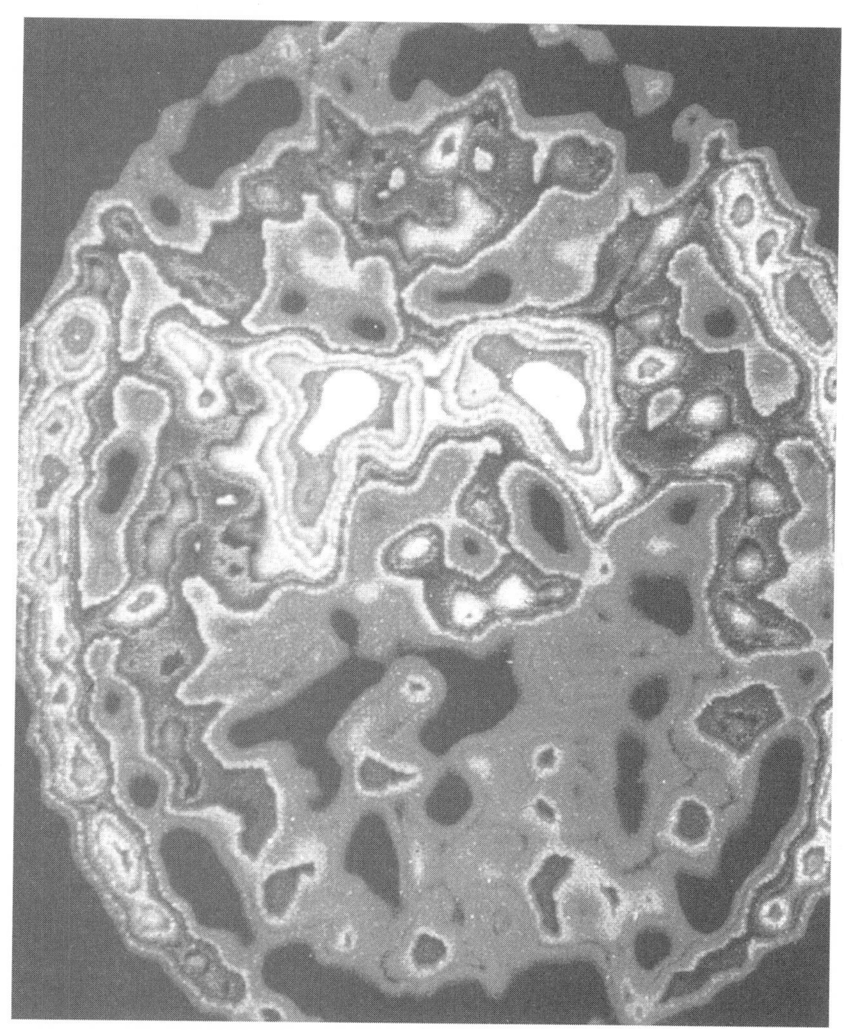

FIGURE 11-4: Imaging a human brain using PET. A small amount of a dopamine precursor labelled with fluorine-18 is injected into the patient. The pattern of the gamma rays detected indicates the distribution of the dopamine, a key chemical involved in brain function. By studying these patterns physicians can diagnose a variety of brain disorders.

brain disturbances such as epilepsy, brain tumours, strokes, and can even give information on the complex workings of the mind.

McMaster University Hospital in Hamilton, Ontario, has one of the major nuclear-medicine departments in Canada. Here, radioisotopes that emit a positron (fluorine-18, carbon-11, nitrogen-13, oxygen-15) are made in a small cyclotron located at the hospital.

Other Nuclear Techniques

The Single Photon Emission Computerized Tomography (SPECT) camera has gained widespread acceptance over the last decade for use in routine clinical imaging. The SPECT camera provides views of thin slices of an organ rather than the entire organ. A detector head rotates 360 degrees around the patient, collecting images that are stored in a computer. Computer-processing techniques are used to reconstruct the slices into three-dimensional images of the organ.

Another instrument, the dual-photon gamma detector, calculates bone densities with great accuracy, which helps immeasurably in the understanding of osteoporosis. The instrument directs gamma rays of two different energies through the bones of interest. By measuring the difference in gamma-ray attenuation, doctors can determine bone density. And if these measurements are repeated over time, doctors can determine how bone density changes with aging.

Studies have shown that strontium-89 can be used to ease pain in patients that have bone cancer (Note 11-3). Not only does the patient feel better, but the treatment costs about $20,000 less than conventional treatment.

Radionuclide Manufacture

MDS Nordion in Ottawa, Ontario, is the world's leading producer of radioisotopes for medical and industrial uses. A high-tech firm spun off from AECL, it supplies about two-thirds of the world demand for reactor-produced isotopes, as well as substantial amounts of isotopes generated in cyclotrons. Isotopes with an excess of neutrons (carbon-14, chlorine-36, nickel-63, molybdenum-99, iodine-125, iodine-131, xenon-133, and others) are produced in the NRU reactor at Chalk River Laboratories.

Thallium-201, widely used in heart studies, is the most significant isotope produced in a cyclotron. A cyclotron, also called a particle accelerator, uses an alternating electric field to accelerate charged particles (such as protons) to a very high speed. They are guided around a circular track by magnets, and then shot at specially prepared targets. The collision produces radioactive substances that are deficient in neutrons including cobalt-57, gallium-67, indium-111, iodine-123, strontium-82, as well as thallium-201. MDS Nordion operates two cyclotrons that are dedicated to radioisotope production; it also has access to the large TRIUMF cyclotron at the University of British Columbia (see Chapter 16).

The impact of these Canadian-produced medical radioisotopes is enormous. About 15 million medical procedures are conducted each year around the world using medical isotopes supplied by MDS Nordion.

The importance of Canada's medical radioisotope production was dramatically demonstrated in late 2007. On November 18, AECL shut down the NRU reactor for planned maintenance. The Canadian Nuclear Safety Commission (CNSC) refused to allow AECL to re-start the NRU reactor; it claimed that AECL had not installed some required safety equipment. An international health crisis ensued as hospitals around the world, now lacking medical radioisotopes—especially Tc-99m—had to cancel or delay medical treatments. Although the health of hundreds of thousands of people was impacted, the CNSC would not relent, nor did AECL comply with installing the safety equipment. To break the impasse, Canada's Parliament unanimously voted to allow NRU to re-start without meeting the CNSC requirement. This was after they had heard evidence from national and international health experts, and had determined the reactor could be operated just as safely as before the shutdown. In addition, the president of the CNSC was fired. The NRU reactor was restarted on December 16, 2007. For more details see Note 11-4.

With the NRU reactor reaching the end of its life, two new 10 megawatt pool-type reactors, called MAPLE 1 and 2 (Multipurpose Applied Physics Lattice Experiment), were to be built at Chalk River. These reactors were to be owned by MDS Nordion (which provided the funding), and operated by AECL. They would produce medical radioisotopes including molybdenum-99, iodine-131, xenon-133, and iodine-125. The project also included a processing facility where the isotopes would be extracted and packaged for shipment to MDS Nordion in Ottawa. However, after more than ten years and an expenditure of several hundred million dollars, the MAPLEs could not be made to perform as required and in May 2008 AECL abandoned the project. In turn, MDS Nordion launched a $1.6 billion lawsuit against AECL for damages and breach of contract.

The MAPLE reactor incident was an enormous debacle, and a black eye for AECL. It is hard to imagine that the organization that engineered the NRU and the CANDU reactors could not cope with the MAPLE system. At the time of writing, the reasons for AECL's failure had not been made public, and it was not clear how the issue would be settled, or how the much-needed medical isotopes will be produced when the aging NRU closes.

Therapeutic Nuclear Medicine

Therapeutic nuclear medicine—which deals with treating and curing diseases—is dramatically different from diagnostic nuclear medicine. In diagnostic procedures, radionuclides are placed inside the body and radiation doses are small. In contrast, the doses administered in most radiotherapy are delivered by sealed sources kept outside the body; the doses delivered are large. Some radiotherapy also involves placing radioisotopes in much smaller doses inside the body; an example of this is the treatment of the thyroid with radioactive iodine.

Cancer is a deadly disease, accounting for 20 to 25 percent of natural deaths in North America. Furthermore, the incidence of some forms of cancer such as lung cancer are increasing by about 0.5 percent per year. Curing cancer is a much sought-after goal.

As long ago as the 1890s, X-rays were used to treat breast cancer. Today, radiation is used to kill cancer tumours by carefully focusing gamma-ray beams generated by either cobalt-60 or accelerators onto the tumour. Cancerous cells are more sensitive to radiation than healthy cells. The beam is aimed at the tumour from many different directions so that nearby healthy cells receive much less radiation than the tumour. It is important that gamma beams of the appropriate energy are used, and that they are focused with extreme accuracy.

Another early application of nuclear medicine was in 1946 when radioactive iodine, placed in an "atomic cocktail," was used to treat thyroid cancer. The patient drank the cocktail and the thyroid gland absorbed the iodine, and the radiation eradicated the cancer cells, curing the patient.

Canada has been a leader in the development of methods for treating cancer. The first cobalt-60 cancer-therapy machine was invented in Canada and was first used at Victoria Hospital in London, Ontario,

Harold Johns (1915–1988) started his career as a professor of physics at the University of Alberta. In 1945, he moved to the University of Saskatchewan, at the same time also working at the Saskatchewan Cancer Commission. It was there that he designed and built Canada's first cobalt therapy machine, one of the first in the world. From 1956 until his retirement in 1980 he was a professor at the University of Toronto. He was inducted into the Canadian Medical Hall of Fame in 1968 and was awarded the Order of Canada in 1976.

on 27 October 1951. For many years, Canada was the only country that could produce cobalt-60; today it supplies over 75 percent of the world's demand.

MDS Nordion has supplied more than 2,500 cobalt therapy machines—over half the number in the world—that are used in hospitals in over 80 countries. These machines deliver over 40,000 treatments each day (Note 11-5). MDS Nordion marketed three different models of radiation therapy units, as well as a sophisticated computer model for planning and tracking radiation treatment. A modern cobalt-60 cancer therapy machine is shown in Figure 11-5. In March 2008 MDS Nordion sold its external beam therapy machine and self-contained irradiator businesses, which will now operate as Best Theratronics Ltd.

Radiation treatment can prolong the lives of patients suffering from cancer for many years. Doctors who specialize in a field of medicine called epidemiology have tracked large numbers of such cancer

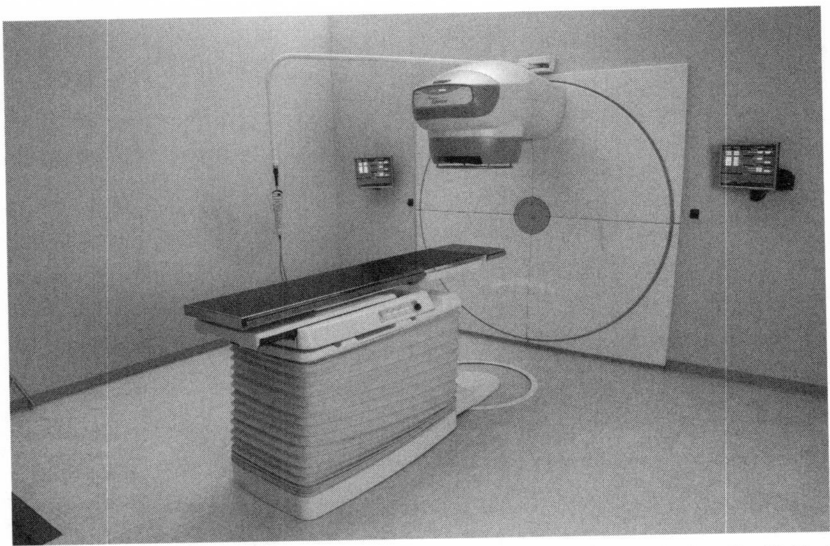

FIGURE 11-5: A cobalt cancer-therapy machine. The cobalt is contained in the heavily-shielded "head" of the machine. The cobalt source is exposed by opening a shutter and the "head" is rotated to aim radiation at the tumour from several different directions to maximize exposure to the tumour while minimizing exposure to the surrounding tissue.

patients; they have calculated that cobalt-60 therapy machines have created approximately 13 million years of additional life.

Another type of radiation therapy known as brachytherapy puts radioactive sources, or "seeds," directly into or near the patient's tumour. This has the benefit of avoiding any doses to healthy skin and tissue elsewhere. These seeds, which are generally produced using cyclotrons, are introduced into the body through surgically placed tubes called catheters.

While the role of radiation in the treatment of cancer is well known, less known is its application to treating heart disease. When coronary arteries become blocked, one option is to use catheterization techniques to open the blockage by installing a stent—a small tube of wire mesh that holds the artery open. The problem is that the stent itself can become blocked. A promising solution to this dilemma is to use a stent containing a special radioactive material that discourages clotting. This method is now coming into increasing use throughout the world.

In conclusion, the impact of nuclear technology on the practice of medicine is profound, and Canada has been a world leader in this field.

-12-

Nuclear Technology in Industry and Science

While the use of nuclear technology in industry and science is less dramatic than it is in medicine, it is even more far-reaching. Almost everything we do and many products we use are of higher quality, cheaper price, or simply offer more choice and variety because of the ingenious use of nuclear technology.

The ability to use radioactive isotopes in so many ways is a result of three unique nuclear properties:

- radiation can penetrate matter;
- radiation can be detected easily and accurately;
- there are thousands of radionuclides in a wide range of radiation energies and activities; they can be made into almost any physical or chemical form.

Irradiation

It was recognized over 130 years ago that germs spread diseases. Since then, various methods of sterilization have evolved, many of them depending on heat. In the 1950s, researchers considered the possibility of using gamma rays for irradiation. This permits the sterilization of products such as plastics that are damaged by heat, and is often the only means of sterilizing some pharmaceutical powders, solutions, and ointments. A particularly important feature is that the penetrating character of gamma rays allows medical supplies to be sterilized after they have been packaged and sealed, thereby preventing subsequent contamination.

In 1958, MDS Nordion's predecessor (AECL's Commercial Products Division) developed the Gammacell 220, the first research irradiator, which was sold to North Carolina University. Today successors of the Gammacell 220—the GC40, GC1000, and GC3000 irradiators—are widely used by hospitals, blood banks, and laboratories for irradiating small samples of blood for immuno-incompetent or immuno-compromised patients, or for research. For example, one of the best ways to reduce the risk of "Graft Versus Host Disease" in patients with severely weakened immune systems—such as premature babies and bone-marrow transplant recipients—is to use irradiated blood when performing transfusions. As noted in the preceding chapter, MDS Nordion sold its external beam therapy machine and self-contained irradiator businesses in 2007, which now operate as Best Theratronics Ltd.

In 1964, the first full-scale commercial irradiator in the world was developed in Canada, and by 2008 there were a total of about 170 irradiation plants in 65 countries around the world. More than 120 of these were supplied to over 45 countries by MDS Nordion.

Today, more than 40 percent of the world's disposable medical products are sterilized with cobalt-60 gamma radiation (Note 12-1). Articles such as syringes, needles, sutures, intravenous tubing, gloves, gowns, sponges, and catheters are packaged and then sterilized while inside the packages. Doctors, scientists, and medical technologists have measured, used, and studied nuclear irradiation for decades on equipment and supplies that have been used to treat millions of patients, and they have reached a simple conclusion: it saves lives by preventing contamination.

Research irradiators are used to study the effects of radiation on a wide variety of materials. For example, electronic components that will be used in satellites are tested in research irradiators to see how they respond to the radiation fields encountered in space.

Nuclear Golf Balls

A sports company in the United States has released "the world's longest legal golf balls." The balls, which are irradiated with gamma rays from a cobalt-60 source, were used to set a world record drive of 367 yards on the fly. The gamma radiation increases cross-linkages between molecules making the golf balls more resilient. The company's line of irradiated tennis strings is also receiving acclaim.

The radioactive cobalt that forms the heart of irradiation units is made in CANDU reactors in Ontario and Quebec and encapsulated by MDS Nordion. The cobalt has only about one-quarter the activity of that in cancer therapy treatment units (see Chapter 11). Some of the stainless-steel adjuster rods used to control the nuclear reaction are replaced with rods made of cobalt-59. These are installed when the reactor is shut down for servicing and are removed after about one year, during which time the cobalt-59 has absorbed neutrons and become cobalt-60. With a half-life of 5.26 years, its radiation intensity decreases by about 1 percent per month. The more active cobalt for cancer therapy is "custom" made in AECL's NRU reactor. Canada provides about 80 percent of the world's supply of cobalt-60.

Food Irradiation

Illness due to contaminated food is "perhaps the most widespread health problem in the contemporary world and an important cause of reduced economic activity," says the United Nation's Joint Expert Committee on Food Safety. And food-borne illnesses, or food poisoning, are not restricted to third-world countries. In Canada, Health Canada and the Public Health Agency of Canada estimate that between 11 and 17 million cases of food-borne illness occur each year with a cost to the health care system of between $12 and $14 billion. The death toll in Canada, based on US statistics, is likely between 200 and 500. Clearly, food irradiation has the potential to alleviate an enormous amount of human suffering.

Gamma rays are sent through a given food item. This disrupts organic processes, breaking down microbial cells such as bacteria, yeasts, and moulds. Parasites and micro-organisms such as trichinella and salmonellae, insects, and their eggs and larvae are killed or made sterile. A commercial food irradiator is shown in Figure 12-1.

One of the world's highest priorities is to produce enough food to feed the ever-increasing population. Considerable effort is invested in fertilizing farmland, improving crop varieties, providing irrigation, and improving animal husbandry. With some countries reporting up to 40 to 50 percent post-harvest losses through infestation of staple

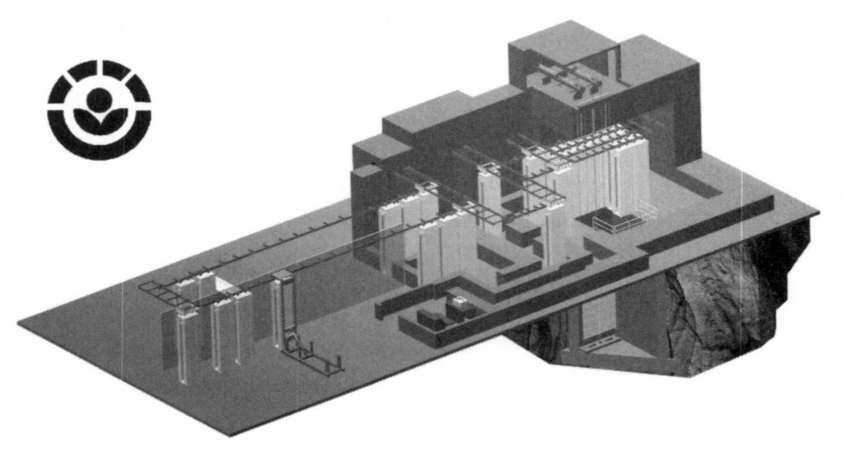

FIGURE 12-1: A full-scale commercial irradiator. These plants irradiate a variety of products including food behind two-metre-thick concrete walls in a bungalow-size building. The cobalt-60 gamma-ray source resides under the concrete building and is raised into position once the packages are in place. Inset is the radura, an international symbol that identifies food that has been irradiated.

foods like grains and yams, it is also important to make progress in preserving food once it has been grown. Preservation methods have evolved from sun drying, to salting, smoking, canning, freezing, heating, and the addition of chemicals. Food irradiation is comparable to pasteurization or freezing and can slow the ripening of some fruit and inhibit the sprouting of root crops. It preserves food by destroying micro-organisms that cause normal spoilage; for example, poultry that has been irradiated will stay fresh for up to two weeks, compared to three days if not irradiated.

Food irradiation has been studied more thoroughly than any other food preservation method. These studies show that irradiation is effective and there are no adverse effects from the consumption of irradiated food. The World Health Organization, the Science Council of Canada, the American Medical Association, and other expert bodies have endorsed food irradiation. It is important to realize that irradiation is virtually identical to the process used by X-ray machines such as at airport security or the dentist's office. Food irradiation does not make food radioactive, just as dental X-rays do not make your mouth radioactive.

Labelling requirements in Canada and the US are similar, requiring irradiated food packages to bear the international radura symbol as well as one of the two statements: "treated with radiation" or "treated by irradiation." The radura is shown inset in Figure 12-1.

Food irradiation is now used relatively widely, especially in developing countries. There are about 170 gamma irradiators worldwide (Note 12-2) and more than 40 countries have approved the use of gamma irradiation for hundreds of foods, including fresh fruits and vegetables, grains, spices, poultry, and seafood. In 1984, the Food and Agriculture Organization and World Health Organization of the United Nations published a general code in respect of food irradiation that states: "The irradiation of foods up to an overall average dose of 10 kilogray introduces no special nutritional or microbiological problems." Based on exhaustive research, this statement forms the basis for food irradiation legislation in many countries.

Interest in food irradiation in the United States was accelerated by various outbreaks of food poisoning. In 1993, for example, four children died and several hundred became ill after eating hamburgers containing the E. coli 0157:H7 micro-organism. In 1997, Hudson Foods recalled 11 million kilograms of red meat including hamburger patties because the meat had been contaminated with harmful E. coli. This industry giant soon filed for bankruptcy. In 2008, a case of salmonella poisoning in tomatoes sickened more than 1,300 people and caused panic across North America with many restaurants refusing to serve tomatoes. The outbreak is estimated to have cost the tomato industry about $300 million.

Irradiation also received a boost because methyl bromide, once a commonly used food pest-control fumigant, was listed as an ozone depleting substance and is being phased out.

Progress is being made in implementing food irradiation in the United States. In 1985, the US Food and Drug Administration approved the irradiation of pork to control the parasite that causes trichinosis. In 1990, it approved the irradiation of poultry, and in 1997, the irradiation of fresh and frozen meats including beef, lamb, and pork. In 1991, North America's first commercial irradiator dedicated to food processing was installed near Tampa, Florida, by MDS Nordion. More recently, progress has been made in pasteurizing meat products using high-energy electrons.

Canada is a leader in developing food irradiation technology. A full-scale food irradiation facility was established at the Canadian Irradiation Centre at Laval, Quebec in May 1987 as a joint venture between University of Quebec's Armand Frappier Institute and MDS Nordion. The Centre is one of the world's leading facilities for research, training, operational demonstrations, and product and market trials of irradiation. A food irradiation research facility has also been established at the Agriculture Canada station at Ste.-Hyacinthe, Quebec.

In contrast to the rest of the world, there is little commercial food irradiation in Canada, in spite of considerable media attention to food poisoning. In 1998, for example, a food poisoning outbreak in Canada caused over 500 people to become ill. Tainted cheese containing the salmonella bacteria came from a single major food processing plant, from where it was shipped all across the country. This large, centralized plant would be well suited to incorporate irradiation, which—as with so many medical products—could take place following packaging.

Foods are normally approved for irradiation on an individual basis by Health Canada. Since 1984, only two food items have been approved: spices and mangos—hardly a staple of Canadian diets. A petition to approve irradiation of poultry, originally submitted to Health Canada in May 1993, was still under review fifteen years later. No meats have been approved. As Health Canada dithers, Canadians continue to get sick and die from food-borne illnesses. It is ironic that the country that is the leader in developing food irradiation technology is the slowest in accepting it.

Inspection and Gauging

Radiography

Just as X-rays are used to look at bones inside humans, radiography uses gamma rays to take "shadow" pictures of industrial components such as aircraft castings and welds in oil and gas pipelines to detect structural faults, impurities, and porosity. Radiography principally

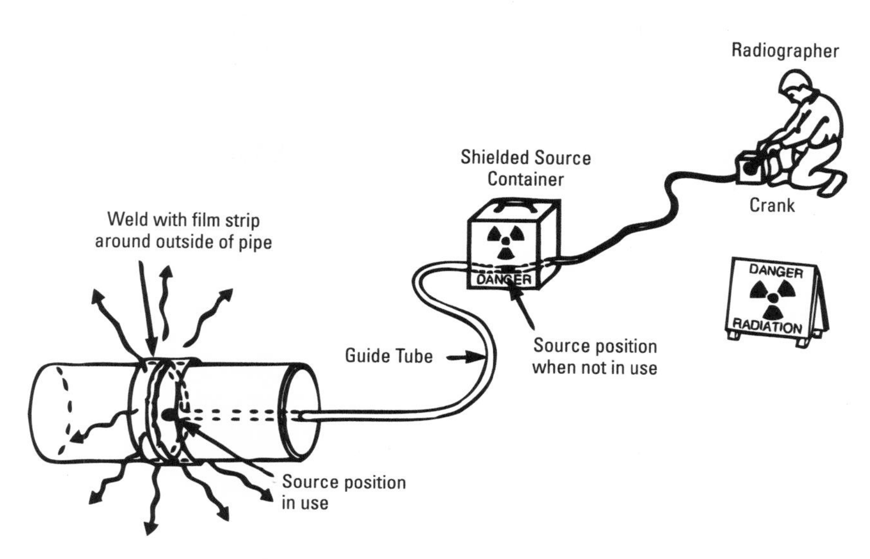

FIGURE 12-2: Crank-out radiography cameras of this type have many applications, such as checking the integrity of welds in oil and gas pipelines.

uses iridium-192 sources (half life of 74.2 days) which are portable, do not require any electrical power, and whose gamma rays are more penetrating than X-rays. More than 800 iridium sources are used in Canada each year, with each one performing 5 to 50 radiographs per day during a useful life of three to six months.

To inspect pipe welds, for example, a crank-out camera is used as shown in Figure 12-2. A shielded container houses a radioactive source, which can be propelled along a guide tube using a crank and long cable by an operator, who remains a safe distance from the container. A strip of film placed around the pipe provides a radiograph of the weld, revealing any defects. More than 140 companies in Canada are licensed to perform radiography. This helps to produce high-quality piping systems and reduces the likelihood of failures and potentially serious consequences such as environmental contamination or explosions.

Nuclear Gauges

Because radiation can be detected and measured with great accuracy, radioisotopes make excellent tracers and gauges. There are two types of nuclear gauges: fixed and portable.

Fixed gauges are most often used in factories to monitor production processes and ensure quality control. A fixed gauge consists of a radioactive source housed within a shielded holder and placed at a critical point in the process. When the shutter in front of the source is opened, a beam of radiation is directed at the material being processed. A detector mounted opposite the source measures the radiation, which is diminished in proportion to the amount and density of material in between.

A significant advantage of nuclear gauges is that they do not need to be in contact with the material being controlled, and can therefore be used on high-speed processes and on materials with extreme temperatures, pressures, or containing harmful chemicals.

Radioisotope gauges for measuring mass per unit area, called thickness gauges, are used in almost every kind of industry in which sheet material is produced (see Figure 12-3). In the paper industry, the accuracy and quick response of these gauges is vital as the paper machines operate at high speed. Similarly, the production of steel plate in modern

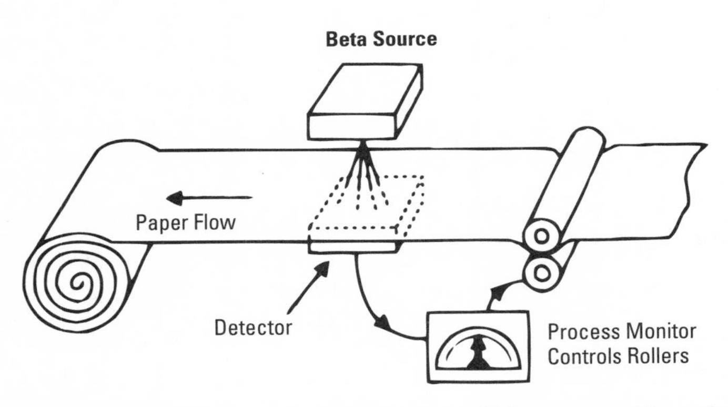

FIGURE 12-3: Schematic of a continuous-process paper-thickness gauge. Many other products such as textiles, sheet metal, and even cigarettes depend on high-speed, high-accuracy radioactive gauges for their production.

rolling mills could not take place without accurate measurement of thickness at every moment of the process and automatic control of the rolling stands. In the plastics industry, radioisotope gauges are used to improve the uniformity of the product. This yields savings in raw material and reduces costs for rejects. Isotopes commonly used for this purpose are cesium-137, krypton-85, and strontium-90.

Another application is the monitoring of material flow in a pipe or chute. In coal-fired electrical-generating stations, for example, the flow of coal into the furnace can be monitored by a radioactive gauge. If a blockage occurs, an alarm sounds, allowing action to be taken to remove the blockage. Comparable methods, based on relating changes in the attenuation of radiation to changes in the materials it passes through, are widely applied to measure the levels in containers and tanks of all kinds.

Nuclear techniques make possible online determination of contaminants such as sulphur and nitrogen (the causes of acid rain) in coal. This method has become routine in the coal industry, with hundreds of millions of tonnes of coal analysed every year. This is clearly essential in minimizing contamination of the atmosphere.

In the galvanizing or tin-coating of steel plate and cans, the exact amount of coating material must be applied: too much coating is expensive, and too little results in early corrosion. Nuclear gauges allow the coating processes to be controlled to tight limits, and up to 10 percent material savings can be realized over other methods.

Portable gauges are used in many industries such as agriculture, construction, and civil engineering to measure moisture in soil and the density of asphalt or compacted clay. There are two basic methods: direct transmission and backscatter.

In direct transmission (which is more accurate), the source is placed in a tube and inserted beneath the surface through a hole created for the purpose. Radiation is transmitted to a detector in the base of the gauge, where it is measured, and the density of the soil calculated based on the amount of radiation detected.

The backscatter method is quicker as it eliminates the need for an access hole, but it is less accurate. Radiation is directed beneath the surface where some of it is reflected or scattered back to the detector in the gauge.

Each nuclear gauge uses one or two small radioactive sources such as cesium-137, americium-241, radium-226, or cobalt-60, placed in a special capsule. This can be as small as the eraser on a pencil or as large as the tube inside a roll of paper towels. The housing that contains the radioactive source is constructed of heavy metal to provide shielding. As these sources contain relatively large amounts of potentially harmful radioactivity, they must be managed carefully and disposed of properly.

Radioisotopes are also used as tracers to find lost buried pipes or leaks in pipes or storage tanks. A short-lived gamma-emitting radioisotope is flushed through the pipes with water while detectors are stationed on the ground above to trace the flow path and reveal any leaks.

Other Applications

Agriculture

For centuries, humans have worked industriously to increase the quantity and quality of food crops. Nuclear technology has made a significant contribution to these efforts. Over the past 50 years, a number of plant breeding programs have been undertaken using mutation induction with radiation to increase disease resistance, improve yields, strengthen stems so the plants can better withstand storms, and increase winter hardiness. Some of these developments have had a significant impact. In Pakistan, for example, a new cotton strain was released in 1983. Cotton production in the country approximately doubled and the crop value of this new type of cotton was more than $2.3 billion in 1988 and 1989.

Water is one of the most important factors in agriculture. Nuclear methods allow continuous monitoring of the moisture content of soil, which helps to determine the amount of water being consumed by plants and evaporation. In this way, soil scientists not only improve crop yields but, in some cases, up to 40 percent of the water can be saved.

Geology

Radioactive sources are routinely used in geophysical well-logging to explore the earth's subsurface. Density gauges—also called gamma-gamma gauges—emit gamma radiation from cobalt-60 or cesium-137 sources and correlate the response to the density of the surrounding rock formation. Density, in turn, is a good indicator of porosity, a key parameter for finding rocks with potential for containing oil or gas. In a similar way, neutron generators are used to determine water content in rock formations because neutrons are absorbed by the hydrogen in water molecules. Radioisotopes are also used as tracers in tracking the connection between oil wells and the progress of secondary and tertiary oil-recovery processes.

Civil Engineering

Neutron and density gauges have been used for about 50 years to make civil engineering, agricultural, and hydrological measurements. Civil engineers, for example, use such gauges to establish how soils in road beds, under building and bridge footings, and in landfills have been compacted and how they will deform under loads.

Insect Control

Many insects have long been a nuisance and threat to human and animal health. Controlling insects with chemicals poses serious problems of environmental pollution. It is also on the whole ineffective, given that insects develop resistance to the insecticides and pesticides. An innovative method of controlling insects that avoids the use of damaging chemicals is the sterile-insect technique. Large quantities of the particular insect are bred and then sterilized using gamma irradiation. When released into the native population, they mate with wild insects but no offspring are produced. This technique—the only environmentally sound method available—is most effective when the sterile insects can be produced in large numbers, and the native insect population is relatively small and isolated from other infestations. The first successful application of this method was used in 1954 to eradicate the screwworm, a devastating pest of domestic animals and wildlife, from the island of Curacao. Later, the screwworm was also eradicated from the United States and then Mexico. Texas ranchers estimated that the program saves them about $150 million annually.

Tsetse flies, which slowly destroy livestock by transmitting a parasitic disease, have prevented the settlement of large areas of Africa. It is estimated that the human form of this disease, sleeping sickness, which eventually causes death, afflicts about 300,000 people in Africa. The tsetse fly has recently been eradicated from the once heavily infested main island of Zanzibar. Tsetse flies were bred in a special fly factory and the male flies were sterilized using low doses of gamma irradiation from cobalt-60 or cesium-137. Almost 8 million sterile flies were dropped by air over infested areas.

This method of eradicating problem insects was developed by entomologists Edward Knipling and Raymond Bushland, who received the World Food Prize in 1992 for the achievement.

Safety

The smoke detector is so effective in saving lives that in many communities it is a legal requirement for them to be installed in new homes. The ionization-chamber smoke detector uses a radioactive material, generally americium-241, to ionize the air between two metal plates. An electric voltage is applied across the plates, generating a small electric current. When smoke enters this space it causes the current to decrease, triggering an alarm.

Tritium (hydrogen-3) is used for lighting at runways, exit signs, and for situations where electrical light could constitute a fire hazard such as in coal mines or grain elevators. Tritium emits radiation that activates phosphors to produce light.

Nuclear techniques also help make airplanes safer. Today's jet engines operate at high temperatures and therefore, as shown in Figure 12-4, they are cast with built-in internal cooling channels. If any of these channels are blocked, the blade may fail due to overheating and cause the whole engine to fail. To ensure the blades are properly made, each one is inspected by neutron radiography using reactor neutrons. This is the only reliable method of detecting blockages in the cooling channels as X-rays are not sufficiently powerful.

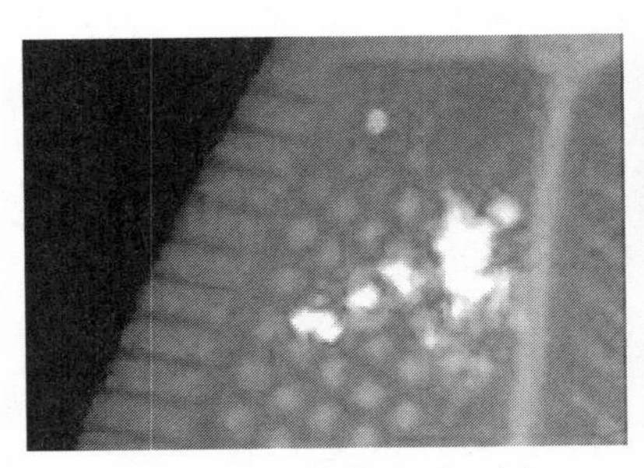
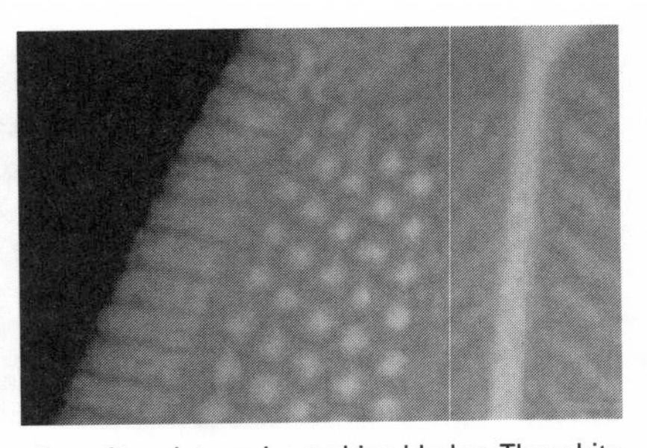

FIGURE 12-4: Neutron radiographs of the same portion of two jet-engine turbine blades. The white traces on the left blade indicate it is defective. The defective blade could overheat and fracture causing the engine to fail.

Detecting Land Mines

Long after the wars in which land mines were used are over, they continue to kill and maim thousands of people every year. A big problem with detecting land mines is discriminating the mines from the many other objects in the ground. For example, tin cans have a similar magnetic signature as a metal land mine and rocks and stones of various types appear on ground penetrating X-rays. Mines made of plastic parts are particularly difficult to identify. Canada is not only a leader in international efforts to ban land mines but is also a leader in research aimed at finding efficient methods of detecting and removing them.

One research program is looking at directing neutron beams into the soil from a specially designed robotic vehicle. Neutrons encountering the high concentrations of nitrogen typical of explosives cause distinct gamma rays to be emitted; these can be detected. This technology is applicable to finding both metal and plastic mines.

Archaeology

Radioactive age dating has become a powerful scientific tool. Carbon-14 (half-life of 5,730 years) is constantly being produced by cosmic radiation striking the nitrogen in the atmosphere. Living plants take up the radioactive carbon as they absorb carbon dioxide. When the plant dies, the uptake ceases and the carbon-14 decays. The longer the plant is dead, the more carbon-14 will decay, and by comparing this amount to normal carbon, the elapsed time since the death of the plant can be calculated. This method has been widely used to date objects containing carbon that are from 1,000 to 40,000 years old, such as soils, shells, marine sediments, trees, archaeological artifacts, bones, and textiles. For example, this method was used to determine the age of the famous Turin Shroud. Similar methods using other radioisotopes have been used to date rocks and unravel the history of the earth.

This chapter has introduced only a few of the ways that nuclear technology as applied to industry, agriculture, and natural resources impacts our daily lives. There are many other examples and more applications are being discovered all the time.

Transportation

The movement of radioactive materials is commonplace, due not only to the many shipments involved in the nuclear fuel cycle, but also to the thousands of radionuclides used in medicine and industry. In Canada, the Canadian Nuclear Safety Commission (CNSC) licenses over 4,000 users of radioactive materials and there are about one million shipments of nuclear materials each year. These shipments are made in accordance with performance standards that are based on international standards developed by the International Atomic Energy Agency (IAEA). These standards are designed to protect the public even in the case of severe accidents. Five different package types can be used, depending on the type of material to be transported and its level of radioactivity. They are described below in ascending order of radioactive content.

1. Excepted packages (also called limited activity packages) are for radioactive material with a very low activity and a very low hazard. Even if released in an accident, the amount of material would not pose a significant hazard. These packages do not require labels or special markings and could be, for example, an industrial-strength cardboard box.
2. Industrial packages contain what are called low-specific-activity materials or surface-contaminated objects such as uranium concentrates and related compounds that are shipped in large quantities in Canada. The hazards, even if materials are released in an accident, are minimal, but there is a potential for nuisance, economic disruption, or environmental degradation. A typical industrial package might be a steel drum marked with a Class 7 "Radioactive" placard.
3. Type-A packages contain radioactive materials of medium activity such as radioisotopes being shipped to hospitals or radioactive sources used in research or in industrial applications. The hazard of an accident is controlled by the amount of radioactivity and the strength of the package. A Type-A package may be a metal,

plywood, or cardboard box or drum with internal foam padding, lead shielding, absorbent material (if a liquid), and other containment features. The package must bear labels and the transport vehicle must be placarded. Ontario Power Generation uses a Type-A Container to ship low-level waste. To obtain a CNSC license, scale modelling and/or computer modelling must demonstrate that the package can withstand a fall of 1.2 metres onto a hard surface and an attempted puncture by a steel bar. If the container is to be used for gas or liquid waste, it must undergo more rigorous testing including a more forceful puncture test and a fall from a nine-metre height.

4. Type-B packages contain radioactive material with large activity such as cobalt therapy sources, radiography cameras, and irradiated nuclear fuel. The hazard of an accident is controlled by the strength and robustness of the packages, which are constructed of steel with depleted uranium or lead as shielding and weigh from 25 kg to 35 tonnes. Type-B packages must be licensed, labelled and the transport vehicles must be placarded.

   Ontario Power Generation has designed several Type-B packages, including a two-bundle cask for irradiated fuel, a bulk-resin flask for ion-exchange resins and higher level waste, and a tritiated heavy-water transportation flask. All Type-B containers must be able to withstand a 9-metre fall onto a hard surface, a 1-metre drop onto a steel spike, 30 minutes in a fire of 800 °C (1,470 °F), and 8 hours immersed below 15 metres of water.

   Sometimes the radioactive material to be shipped in Type-A or -B packages is placed into a special capsule or is put into a ceramic-like form to provide additional containment and safety.

5. When fissile materials are shipped, the word "Fissile" or the letter "F" is added to the description, as in a Type-AF package.

   The labels placed on these packages must indicate the maximum radiation level at the surface of the package and at a distance of 1 metre. Transportation regulations and practices in most countries are similar as they are based on IAEA guidelines.

-13-

Uranium: The Nuclear Fuel

Uranium is a fascinating element. In order to fully understand nuclear power, uranium itself must also be understood as it is the fuel that drives reactors. The entire process, or life cycle, through which uranium passes in generating electricity, is depicted in Figure 13-1.

The nuclear fuel cycle in Canada begins with the mining of uranium ore from the ground. The ore goes through a complex milling, refining, and conversion process and then it is manufactured into nuclear fuel bundles. The bundles are loaded into CANDU reactors where the heat produced by nuclear fission is used to generate electricity.

Although the mining of uranium ore and its subsequent transformation into fuel is a complex process, the cost of nuclear fuel over the lifetime of a reactor is significantly less than for an equivalent coal or natural-gas plant. This is because a kilogram of uranium fuel generates about 60,000 times more energy than a kilogram of coal, so far less nuclear fuel is required. Furthermore, other impacts of mining and preparing nuclear fuel such as energy consumption are less than for the fossil fuels, especially coal.

After the fuel is consumed, it is removed from the reactor and stored on-site for a number of years while its radioactivity and heat subside. The feasibility of centralized long-term spent fuel storage is also being assessed. At some point in the future the spent fuel will be encapsulated in sturdy, leach-resistant containers and placed in permanent disposal by burying it deep underground from where it originated, thereby completing the cycle.

The nuclear fuel cycle for light-water reactors is different from the Canadian approach as it includes enrichment of the uranium-235 in the fuel. The French, Japanese, Russian, and British fuel cycles differ again by including reprocessing of their spent fuel. This involves the removal of some fissile elements from the spent fuel so they can be

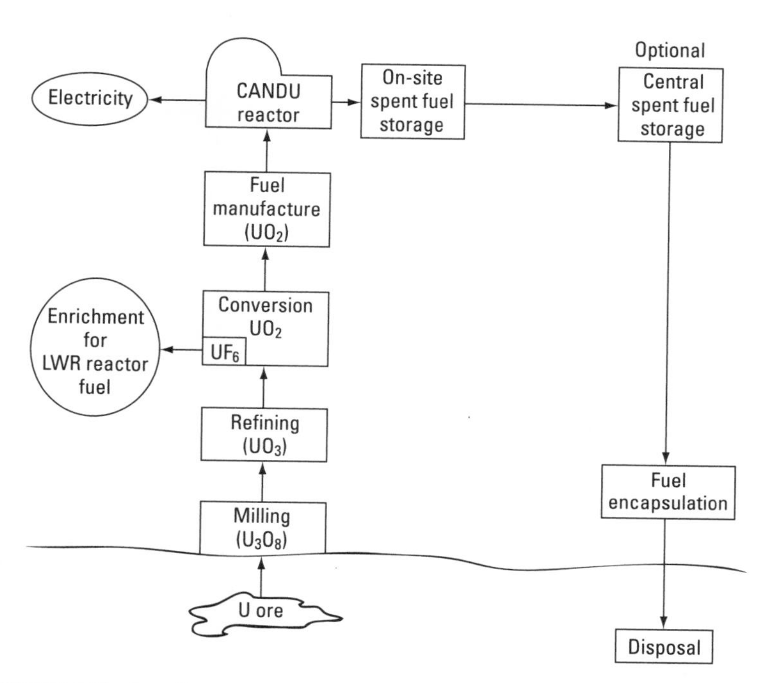

FIGURE 13-1: The Canadian uranium life cycle from mining of uranium ore through preparation of fuel to creation of electricity in CANDU reactors to final disposal deep underground.

incorporated into fresh reactor fuel. They do not plan to have long-term central spent-fuel storage.

Some of the steps comprising the nuclear fuel cycle are described in earlier chapters. Reactors and how they operate are described in Chapters 6 and 7; waste management, including interim on-site storage, long-term central storage, and permanent disposal, is described in Chapter 10. The remaining components of the nuclear fuel cycle—mining, refining/conversion, fuel manufacture, reprocessing, and transportation—are described in this chapter.

Uranium Mining

Uranium was discovered in 1789 by Martin Klaproth, who named it after the recently discovered planet Uranus. Interestingly, Klaproth also discovered zirconium—another important element in the nuclear field. Uranium is one of the more common heavy elements in nature, 500 times more abundant than gold and about twice as common as tin. It is present in virtually all rocks and soils as well as in rivers and oceans. Traces are found in food and human tissue. Granite, which makes up about 60 percent of the earth's crust, averages about four parts per million (ppm) uranium, although the concentration is highly variable. Phosphate rock used to produce fertilizer can contain as much as 400 ppm uranium, and some coal deposits contain up to 1,000 ppm. Long used to add colour to glass, ceramics, and porcelain dentures, it is only in the past half century that uranium has become a valuable energy source. Uranium deposits with concentrations of about 1,000 ppm and greater of uranium may be considered "ore," that is, they may be economic to mine.

Prospecting, the search for valuable minerals holds a romantic place in the history of Canada. Early discoveries of uranium were made in the traditional method by grizzled prospectors braving the wilderness on foot and canoe. But modern times have led to modern methods. Today, aircraft specially equipped with radiation detectors fly low-level surveys over areas that, based on geological maps, appear promising. Once an area has been identified as having potential, field crews investigate it on the ground using hand-held radiation detectors such as Geiger-Muller counters (see Appendix A). Water, soil, and vegetation samples are also collected and analyzed for their uranium content.

Considerable detective work is required to find uranium deposits. For example, if uranium-bearing boulders are discovered, geologists will reconstruct the movement of past glaciers to determine from where they may have come. If uranium deposits are hidden deep underground, geophysical methods such as electrical-resistivity surveys are conducted to find materials like graphite, which are good conductors of electricity and are often associated with uranium deposits. Once the search has been narrowed to a small target area, drill rigs are brought in, often by float-planes, and extract rock cores from the subsurface. Chemical analyses of the samples determine whether uranium is present in economic concentrations and help define the size and shape of the ore body.

Canada has been a major world producer of uranium since the first global demand for this material. Today, the only producing area is northern Saskatchewan, although other areas have been active in the past. Canada is the world's leading exporter of uranium and hosts

Gilbert Labine (1890–1977) was born in Westmeath, Ontario, and attended the Provincial School of Mines at Haileybury, Ontario. He became a prospector and mine developer founding Eldorado Gold Mines. In 1930, he discovered Canada's first major uranium deposit at Port Radium on Great Bear Lake in the Northwest Territories (where this picture was taken in 1935). Eldorado was taken over by the government of Canada in 1944, and Labine turned to other successful mining ventures. At the time of his death he was a respected and wealthy senior statesman in the mining community.

three of the top ten producing mines in the world. To give a different perspective: Canada's production of 11,180 tonnes of uranium oxide (U_3O_8) in 2007 contained more than twice the energy available from Canada's total annual oil production. Total world production of uranium oxide that year was 48,680 tonnes, with Canada producing 23 percent of the global supply.

The story of Canada's uranium mining industry began in 1931 in the cold and forbidding north, when prospector Gilbert Labine discovered "pitchblende," a uranium-bearing mineral, near the shores of Great Bear Lake in the Northwest Territories. His discovery resulted in the development of a mine at Port Radium in 1932 and a refinery in Port Hope, Ontario, in 1933, both owned and operated by Eldorado Gold Mining Company. The original objective was to produce the rare and precious element, radium, which is found in uranium ore, and was felt to be a miracle cure for cancer. It commanded prices as high as $75,000 per ounce until the market for this crashed in the late 1930s.

During World War II, the demand for uranium took centre stage as the United States and its allies, Britain and Canada, began the Manhattan Project to develop the first nuclear weapon. The Canadian government took over Eldorado and formed a Crown corporation, which was later renamed Eldorado Nuclear Limited, and the Port Radium mine was re-opened. The demand for uranium for weapons production continued in the post-war era. Port Radium produced uranium until the mine was closed in 1960. By the early 1960s, demand began to be driven by the development of nuclear power for the production of electricity.

Ontario has also seen considerable uranium activity. The Bancroft area witnessed radium mining in the 1920s and 1930s, and lived through two uranium booms from 1956 to 1964 and from 1976 to 1982. Faraday, Bicroft, and Madawaska Mines, all abandoned today, produced about 6,700 tonnes of uranium oxide (U_3O_8) using underground mining methods.

Elliot Lake, Ontario became a uranium boom-town virtually overnight. In 1954, Denison Mining intersected uranium in exploration drilling, and a brief three years later the first mine was in production. From there, Elliot Lake grew rapidly and soon gained a reputation as the "Uranium Capital" of the world. Two major companies, Denison Mines Ltd. and Rio Algom Ltd., operated 12 mines and their accompanying mills. The ore, with a grade of approximately 0.1 to 0.2 percent uranium oxide, was mined from a depth of about 170 to 950 metres using underground mining methods.

As a result of the continuing foreign military demand, Canada's uranium mining industry continued to grow until 1959, when more than 12,000 tonnes of uranium oxide were produced, yielding $330 million in export revenue—more than any other mineral. Over the next few years, however, the military demand declined and the number of mines operating in Canada decreased to four. Uranium exploration waned and in 1965, Canada implemented a policy of selling uranium for peaceful purposes only. Decreasing demand led to the Canadian government stockpiling uranium until 1974 to support the industry. Thereafter, the uranium industry again experienced growth, a result of the demand for electricity-generating nuclear reactors.

After three decades as the uranium capital of the world, Elliot Lake bowed to the inevitable fate of all mining centres. Unable to withstand the strong competition from the much higher-grade ore bodies in Saskatchewan and their lower production costs, the Ontario mines were decommissioned in the early to mid 1990s, following the production of over 550,000 tonnes of U_3O_8. In 1996, with the closure of Stanleigh Mine, Saskatchewan became the sole province producing uranium.

Northern Saskatchewan

The Athabasca Basin of northern Saskatchewan has been the site of all major Canadian uranium discoveries in the past 40 years. The first northern Saskatchewan uranium deposits were discovered in the early 1950s and Eldorado began mining at Beaverlodge Mine in 1953.

In 1968, the Rabbit Lake deposit was discovered in northern Saskatchewan by Gulf Minerals Ltd. and the German-owned Uranerz Exploration and Mining Limited. By 1975 the mine and mill were in operation. Rabbit Lake was subsequently sold to Eldorado Nuclear (now Cameco Corporation) in 1981. The deposits at Rabbit Lake include the mined-out original Rabbit Lake open pit, the mined-out Collins Bay open pits, as well as the currently operating Eagle Point underground mine.

In 1975, the French-owned company Amok Ltd. discovered the Cluff Lake deposit, which operated from 1980 until 2002. The year 1975 also marked the discovery of the large Key Lake deposit by Uranerz Canada Ltd. A 50 percent interest in this deposit was sold to Saskatchewan Mining Development Corporation, a provincial Crown corporation. Production began in 1983, and Key Lake became the highest grade and largest uranium mine in the world at the time. Mining ceased in 1997.

Mining in the 1980s and 1990s was primarily by the open-pit method as the deposits were near the surface. Surface mining is more economical than underground mining and, combined with the very high ore grade found in the Athabasca Basin, made this uranium very competitive in world markets. The high ore grade also requires that great care be taken to ensure radiation protection for workers.

In 1988, the federal and Saskatchewan governments agreed to the amalgamation of their respective crown corporations, Eldorado

Table 13-1 Summary of Saskatchewan Uranium Mines

Mine	Reserves*	Grade**	Owner	Dates of Operation
Beaverlodge			Eldorado	1953–1982
Rabbit Lake	782	1.3	Cameco	1975–
Cluff Lake	20		AREVA	1980–2003
Key Lake	131	0.41	Cameco	1983–2001
McClean Lake	23	3	AREVA	1999–
Cigar Lake	577	18	Cameco	2005– ***
McArthur River	845	21	Cameco, AREVA	2000–

* 1,000 tonnes ore
** % U_2O_3
*** not producing in 2008

Nuclear Limited and Saskatchewan Mining Development Corporation. The resulting company, Cameco Corporation, is currently the world's largest uranium producer, having purchased most of Uranerz's Canadian holdings. In 1987, Cogema Resources Inc., a French government-owned company, purchased all of Amok's Canadian interests as well as some of the Uranerz holdings. In 2006 Cogema was renamed AREVA Resources Canada Inc. AREVA is the other major player in Saskatchewan uranium mining; it also produces uranium in other countries, ranking third in total uranium production.

The Cigar Lake deposit was discovered in 1981 and construction started in the summer of 1997. The Cigar Lake ore body is the richest uranium deposit in the world, with an average ore grade of 18 percent U_3O_8. It is also one of the largest, with geological reserves totaling 103,000 tonnes of U_3O_8. It will be mined from tunnels above and below the ore zone, using a water-jet boring technique on ore after is has been frozen. The ore will be crushed, ground, mixed with water, and then pumped as slurry to the surface for transportation to the mill. Special remote-control mining methods will be necessary, given the high radiation fields. However, as the result of a rock fall, the Cigar Lake mine flooded during construction in October 2006. The process of pumping out the water started in mid 2008 and production is now scheduled to start by 2011.

In 1988, the massive, high-grade McArthur River ore body was discovered. With geological reserves totalling 158,000 tonnes of U_3O_8 at an average grade of close to 20 percent U_3O_8, this is the largest known high-grade uranium deposit in the world. The McArthur River mine commenced production in July 1999 and is operated by Cameco Corporation. At a depth of 550 metres, it is mined by underground methods similar to those to be used at Cigar Lake. The ore is processed at the Key Lake mill and the mill tailings, or wastes, are placed into the former Deilmann pit of the Key Lake mine, which has been converted into a tailings management facility. The facility uses the "pervious surround" method, which has been successfully used at the Rabbit Lake mill since 1985. Mining at McArthur River produced 8,500 tonnes of U_3O_8 in 2007.

As noted Canada's uranium production in 2007 was 11,180 tonnes U_3O_8, which represented about 23 percent of world output. About 75 percent of this production came from the McArthur River mine alone. Canada's identified uranium resources are about 9 percent of the world total with Australia's identified uranium resources about three times as large.

With their enormous reserves and exceptionally high uranium concentrations—about 100 times the world average grade—the Cigar Lake and McArthur River mines are the future of Saskatchewan uranium mining. They eclipse any other deposits in the world.

The enormous bounty of uranium buried in the Athabasca Basin is almost beyond reckoning and can provide substantial wealth to the province of Saskatchewan and Canada for decades to come. It is estimated that the energy contained in these deposits is equivalent to 17 billion barrels of oil or at a typical mid-2008 oil price of $130 per barrel about $2.2 trillion.

International

Virtually all of the world's uranium is supplied by eight countries: Canada, Australia, Kazakhstan, Niger, the Russian Federation, Namibia, Uzbekistan, and the United States.

In Europe and Japan, limited quantities of reactor fuel are obtained by reprocessing spent fuel. It is estimated that reprocessing will supply less than 5 percent of the western world's uranium requirements in the next decade.

In 1994, the United States and Russia agreed to convert highly enriched uranium from dismantled nuclear weapons into low-enriched uranium for use in nuclear power plants under the Megatons to Megawatts program. By 2006, Russia had converted 250 tonnes of highly enriched uranium from weapons (more than half the committed amount) to low enriched uranium suitable for US light-water reactors. The program will end in 2013.

Uranium demand for existing reactors is easy to predict because it depends solely on the number and size of nuclear reactors, information that is well known. With the high price of oil, however, there has been renewed interest in nuclear power, and the number of new reactors that may be committed in the next several years is difficult to estimate. World uranium production has grown from 42,200 tonnes of uranium oxide in 1996 to 48,680 tonnes in 2007. The western-world electric utilities accumulated large inventories of uranium during the late 1970s and 1980s, but these inventories were depleted by 2000. Even with uranium from the dismantlement of nuclear weapons, the production from existing mines may not be sufficient to meet demand. Without the new mines that are being developed in Canada and elsewhere, there could be a shortfall in the next decade.

The prospect of a potential uranium shortage has set off a small uranium exploration boom with increased levels of prospecting both in Canada and abroad. Reflecting the same fears, the price of uranium has risen sharply, but is relatively volatile. For example, in 2003 the price of a kilogram of natural U_3O_8 was about US$22; by mid 2007 it was almost US$300 per kilogram. By mid 2008, it was back down to around US $150.

Health and Safety

The greatest risk associated with uranium mining is the same as is faced in all mining operations: industrial accidents. Safety in uranium mines is further complicated by the presence of the radioactive gas radon and its daughter products, which can irradiate lung tissue when inhaled. This is a particular concern in underground mining because of the confined spaces. The potential hazard can be minimized by using powerful ventilation systems in underground mine spaces, special respirators in some cases, and limiting work time in some areas. All workers carry radiation monitors so that the radiation doses they receive are measured and tracked. The Canadian Nuclear Safety

Commission (CNSC) sets limits for worker doses and monitors the mines to ensure they are in compliance.

Export Policies

Canada has several policies that govern the export of uranium and nuclear technology. Administered by the CNSC, these are designed to increase the benefits to Canada of developing its uranium resources and to prevent these resources and technologies from being used to manufacture nuclear weapons.

Federal guidelines state that at least a 30-year reserve of uranium supply must be available for all operating reactors in Canada, as well as for those planned or committed for the next ten years. In addition, sufficient uranium production capacity must be available for the domestic nuclear program to reach its full potential. Only after these requirements are met can uranium be exported.

Uranium Processing

Once the uranium ore is extracted from the ground, it must undergo a number of chemical processes before it is ready to be made into fuel for nuclear reactors.

Milling

Uranium ore is first treated in a mill at or near the mine site. The ore is ground to a very small size and then the uranium is leached from the rock, usually using sulphuric acid, and dissolved. The solution is extracted and dried, or calcined, to yield a yellow powder called "yellowcake." Yellowcake contains about 70 to 90 percent U_3O_8 by weight in addition to several other uranium compounds. The yellowcake is then shipped to a refinery.

As is the case for all mining operations, large amounts of waste called tailings are created during the mining and milling process. These are a mixture of solids and liquids called a slurry. The slurry is acidic from the chemicals used in the milling process, so lime or limestone is added to neutralize it. The slurry is then discharged into a tailings management area—effectively a large pond—where solids precipitate out and water is drained off. The solid tailings that remain are similar in composition and hazard to the ore that was originally mined. They are, however, more mobile and must be properly contained.

Uranium tailings contain radium and other hazardous materials. In 2007, there were about 216 million tonnes of uranium tailings in Canada, of which 176 million are in Ontario and about 40 million in Saskatchewan. The amount in Saskatchewan grows steadily as mining continues. Uranium mines account for about 2 percent of all accumulated mine tailings in Canada.

At Elliot Lake, large amounts of mine tailings were created because of the low grade of the uranium ore. These were placed into nearby natural depressions and lakes and were contained by dams to form tailings or impoundment areas. In the early days these tailings ponds were not always designed to currently acceptable standards, although these concerns have been addressed. The runoff water is treated in a series of ponds where barium chloride is added to precipitate the radium still left in the water. The ponds are carefully monitored and after treatment the water, which must meet regulatory standards, is released into the natural waterways of the area.

More advanced tailings-management methods are being used in Saskatchewan to minimize environmental impact. At the Key Lake mill, the mined-out Deilmann open pit is now used as a tailings facility. Instead of using traditional impermeable liners like bentonite or plastics, the pit is lined with separate layers of sand and gravel, which are highly permeable to the flow of groundwater. The method takes advantage of the relatively permeable nature of the blast-fractured pit walls compared to the surrounding rock. De-watered tailings are placed into the pit and compacted. Seepage from compaction of the tailings is pumped from tunnels below the mine into the mill for reuse. The compacted, more dense, and less permeable tailings are surrounded by a less dense, more permeable envelope, through which groundwater will flow, thus bypassing the waste tailings and leaving them undisturbed. A lake will be allowed to form over the top of the tailings area, eliminating the risk of airborne radioactive dust, reducing radon emissions, and avoiding inadvertent human intrusion. The Key Lake tailings management facilty is shown in Figure 13-2. This

FIGURE 13-2: The Key Lake tailings pond is a modern facility.

"permeable-surround" or "hydraulic-bypass" concept is also to be used for disposal of tailings from the Rabbit Lake and McClean Lake mills.

In earlier mining operations such as at Bancroft, however, tailings were generally not disposed of in as rigorous a manner and have left some environmental problems; these problems are now being addressed.

Refining

Yellowcake must be chemically refined to separate the uranium from the remaining impurities and convert it to uranium trioxide. This process was originally done at Port Hope, Ontario, but since 1983 has taken place at Cameco's facilities at Blind River, Ontario, the world's largest and Canada's only uranium refining facility. Refining is a complex chemical process involving the following steps:

1. dissolution of yellowcake in nitric acid resulting in a uranyl nitrate solution
2. impurities are removed from the uranyl nitrate solution
3. water is evaporated from the solution
4. the uranyl nitrate is heated to form uranium trioxide powder

Chemicals are recycled at every step to ensure extraction of virtually all of the uranium and to reduce wastes. The nitrogen oxides produced in step 4, for example, are converted to nitric acid, which is then used again in step 1.

Finally, the pure uranium trioxide powder is shipped to Port Hope for the next processing step, called conversion.

FIGURE 13-3: Aerial view of Cameco's uranium refinery at Port Hope, Ontario. The plant performs the "conversion" step to produce uranium dioxide for CANDU reactor fuel and uranium hexafluoride to feed the uranium enrichment facilities in other countries.

Conversion

Cameco operates Canada's only conversion plant at Port Hope, Ontario (Figure 13-3), where two main products, uranium dioxide (UO_2) and uranium hexafluoride (UF_6), are produced.

About 20 percent of the uranium received at Port Hope is made into ceramic-grade UO_2 powder. A complex chemical process is used that starts by dissolving UO_3 from the refinery in nitric acid. The main waste generated in the conversion to UO_2 is liquid ammonium nitrate, which is concentrated and sold as a fertilizer to local farmers. This "nuclear" fertilizer contains less uranium and radium than many commercial fertilizers. The final UO_2, a brown powder, is shipped to Canadian manufacturers who make the fuel bundles used in CANDU reactors.

The remaining 80 percent is made into UF_6 for subsequent enrichment outside Canada; this is used in light-water reactors. To produce UF_6, the UO_3 from the refinery is pulverized to a fine powder and then reduced with hydrogen into UO_2. In the next step, the UO_2 is mixed with hydrofluoric acid (HF) to make uranium tetrafluoride (UF_4). In the final step, the UF_4 is burned with fluorine gas to produce gaseous UF_6, which condenses to a solid white crystalline form as it cools. It is subsequently heated to a liquid for transfer into steel shipping cylinders where it is allowed to cool into a solid form, prior to shipment to enrichment plants in Japan, Europe, and the United States.

Emissions from the UF_6 plant pass through an acid-recovery system and then a scrubber to capture and reduce fluoride, which is harmful to vegetation. The scrubbing solutions are treated to remove uranium and then lime is added to create calcium fluoride, the only waste from the UF_6 process. In past years this waste was discarded in landfills, but recently this material has been recycled to steel mills, where the calcium fluoride is used as a flux to separate impurities from steel.

In 2007 it was discovered that there had been some leakage of uranium and arsenic/fluoride byproducts from the UF_6 plant into the underlying soil. This had been taking place over a number of years. Groundwater flowing through the soil may have carried some of this contamination into Port Hope harbour. The plant was closed for the removal of the contaminated soil and this was ongoing in mid-2008. This situation arose during a period of concern raised by some Port

Hope residents about the continuing presence of the conversion plant on the Port Hope waterfront.

Fuel Fabrication

Zircatec Precision Industries Inc. (owned by Cameco since 2006) at Port Hope, Ontario, and General Electric Canada (now GE Hitachi) at Peterborough, Ontario, manufacture fuel bundles for CANDU reactors (see Figure 13-4). The first step is sintering uranium dioxide in a hydrogen atmosphere at about 1,800°C into ceramic pellets about one centimetre in diameter and a length of about 2 centimetres. Thirty pellets are assembled into a long tube, known as a fuel element, made of an alloy of the metal zirconium. Elements are assembled into fuel bundles, which are about 50 centimetres long, about 10 centimetres in diameter, and weigh about 22.5 kilograms. The fuel bundles are designed to allow the flow of water through and around them. The fuel contains few impurities that can absorb neutrons. Alloys of zirconium are used to contain the fuel pellets because they are very resistant to corrosion, and absorb very few neutrons compared to stainless steel.

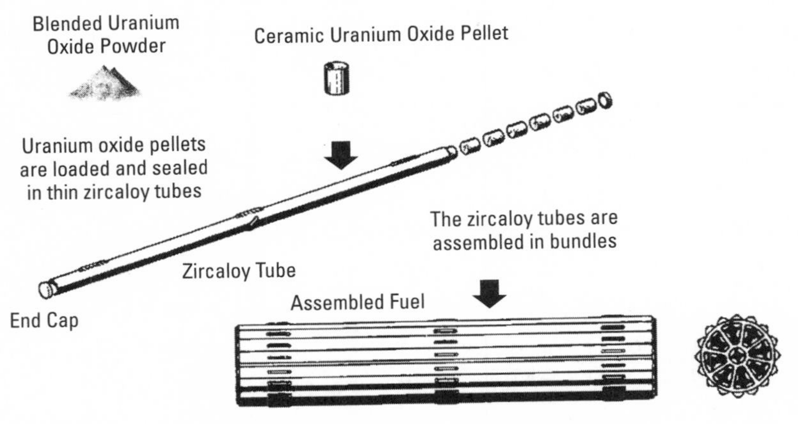

FIGURE 13-4: Schematic illustrating the steps in CANDU fuel fabrication. Uranium oxide powder is compressed to form ceramic pellets about 2 centimetres in length and 1 centimetre in diameter. The pellets are sealed into zirconium alloy tubes which are then welded together to form a fuel bundle.

Uranium Enrichment

Approximately eighty percent of the uranium received at Port Hope is made into uranium hexafluoride and shipped to enrichment plants outside Canada. There are two main enrichment technologies currently in use, both of which enrich the concentration of uranium-235 to between 3 to 5 percent as required in light-water reactors. No enrichment is conducted in Canada. The United States, France, Britain, and Russia operate gaseous diffusion enrichment plants; uranium hexafluoride is heated until it is a gas and then is pumped repeatedly through porous metal filters with holes one-millionth of a centimetre in diameter. Uranium-235, which is very slightly lighter than uranium-238 because of its three fewer neutrons, passes through the filters marginally faster than the uranium-238. In this way, the gas on one side of a filter becomes enriched. By passing the gas through thousands of filters the concentration of uranium-235 is continuously enriched. Once complete, the enriched UF_6 is shipped to fuel manufacturers who convert it back to UO_2 and process it into fuel pellets and fuel bundles of the shape required for light-water reactors.

The gaseous diffusion process consumes large amounts of energy and is being superseded by the gas-centrifuge process that consumes only about 4 percent of the energy of an equivalent gas diffusion plant. Centrifuge enrichment plants are now operating in Britain, the Netherlands, and Russia and by 2008 were responsible for more than half of the world's enrichment. The UF_6 is heated until it is in gaseous form; it is then spun in a centrifuge at very high speeds. The heavier uranium-238 atoms move to the outside of the centrifuge, leaving the central parts slightly enriched in the lighter uranium-235 atoms. By repeating this process many times, the uranium-235 concentration is increased to the desired level.

While still in the experimental stage, laser enrichment appears to be a promising technology. Since uranium-235 and uranium-238 absorb different wavelengths of light, different frequencies of laser can be used to excite and separate the two isotopes. The SILEX process developed in Australia appears sufficiently promising that in 2008 Cameco bought a share of the company that is developing it.

Reprocessing

Reprocessing is to nuclear waste what recycling is to regular garbage: it removes some useable (fissile) materials from spent fuel that can be reused to generate more energy. This extends fuel supplies and reduces the radioactivity and quantity of waste to be disposed.

As fuel is burned in a CANDU reactor, some—but not all—of the uranium-235 is depleted. At the same time, some plutonium-239, a fissile material, is created. The quantity of fissile material in irradiated fuel is only decreased by about 30 percent from that in the fresh fuel. Irradiated fuel is removed from the reactor, not because there is no more energy left to extract, but because too many neutron-absorbing fission products have built up (see Table 10-1).

Instead of mining more uranium from the ground, the plutonium and uranium-235 could be "mined," that is, reprocessed, from the spent fuel and reused to make new fuel and generate more power.

Canada has never undertaken commercial-scale reprocessing and has no plans to do so in the foreseeable future, given the high costs involved as well as the concern that plutonium could be diverted and manufactured into nuclear weapons. The United States has been reprocessing spent fuel since the 1940s as part of their weapons program; but no reprocessing of civilian power-reactor fuel has been undertaken since 1977. The Bush administration reversed this policy in 2005 and efforts are now underway to put into place proliferation-resistant reprocessing facilities in the United States.

Reprocessing is a complex chemical procedure that involves chopping up spent-fuel elements and dissolving the pieces in nitric acid. The uranium and plutonium are then chemically separated. These steps must take place by remote control in shielded rooms; the spent fuel, of course, is highly radioactive. The extracted uranium is returned to the conversion step prior to fuel fabrication; the plutonium goes straight to the fuel fabrication plant.

Reprocessing forms an integral part of the nuclear fuel cycles of some countries like the United Kingdom and France. There are several reprocessing plants worldwide with a total capacity of about 5,000 tonnes per year. The United Kingdom has two plants: an older 1,500-tonnes-per-year plant has been in operation for over 40 years at Sellafield, primarily for metal fuel elements; a newer thermal oxide reprocessing plant (THORP) was commissioned in 1994. At La Hague, France, two 800-tonne-per-year plants are operating and two more are planned. In addition, India now has three plants of total capacity 260 tonnes per year, and a large future increase in its national reprocessing capacity is projected. Russia has a 400-tonne-per-year plant. Japan operates a new 800-tonne-per-year plant. Countries like Japan are turning to reprocessing because they lack indigenous fuel sources and they wish to be energy independent. About 90,000 tonnes of spent fuel from electrical power reactors had been reprocessed by 2005.

-14-

The Fission Future

Nuclear energy today is at a cross-roads. On the one hand, it can be argued that the use of nuclear power should increase because it is reliable and environmentally clean and produces no global-warming emissions.

On the other hand, there is a distinct anti-nuclear movement and in some countries such as Sweden and Germany, and governments have mandated that their nuclear power reactors be phased out.

Many countries, however, particularly those on the Pacific Rim such as Korea, Japan, and China, are vigorously expanding their nuclear programs. And although no new reactors have been built in the United States since the early 1990s, several new reactors are now planned and a few have already been ordered.

In Canada, the province of Ontario has announced that it wants between two and four new reactors. In March 2008, a process was initiated to select a vendor and a reactor type. A significant change from the past is that Ontario is considering other reactor types in addition to the CANDU. Bids are expected from Atomic Energy of Canada Limited (AECL), AREVA, and Westinghouse. The first two of these reactors will be constructed at OPG's Darlington site.

Bruce Power has also conducted studies into a possible Bruce C plant, which could host up to four new 1,000-megawatt reactors. An application for a licence to prepare the site for a new plant was submitted to regulators in August 2006. In November 2008, Bruce Power announced it was also looking at a site in Nanticoke on the shores of Lake Erie as a possible site for a new nuclear plant. The Ontario government indicated that it does not support this proposal.

Elsewhere in Canada, New Brunswick is considering adding a second reactor at the Point Lepreau site. In 2008, the Alberta government was developing a policy on nuclear power with future reactor construction in mind. This action was catalyzed by Bruce Power Alberta, a subsidiary of Bruce Power, which announced it is looking to construct up to four reactors in the Peace River district. In 2008, Bruce Power also initiated a feasibility study of power reactors for Saskatchewan.

A complex combination of social, political, technical, economic, and emotional factors will determine the future of nuclear power. In making such decisions we need to look beyond the status of nuclear power as it stands at the present moment and also consider how it might evolve in the future.

Nuclear power, more than any other power source, involves high technology, an area where bright minds are at work and new ideas are constantly arising. Reactor design is always evolving and improving. Let us look at what advances are being pursued that might have an impact on decisions regarding the future of nuclear power.

The Generation IV Process

The Generation IV International Forum (GIF) was established in January 2000 to investigate innovative nuclear reactor concepts for meeting future energy challenges. GIF members include Argentina, Brazil, Canada, China, European Atomic Energy Community, France, Japan, Russia, South Africa, South Korea, Switzerland, United Kingdom, and United States, with the OECD-Nuclear Energy Agency and the International Atomic Energy Agency as observers.

Figure 14-1 shows reactors grouped in "generations" reflecting their stages of evolution. Generation I consists of early prototype reactors; Generation II comprises the first wave of commercial reactors; and Generation III reactors are the most recent commercial reactors.

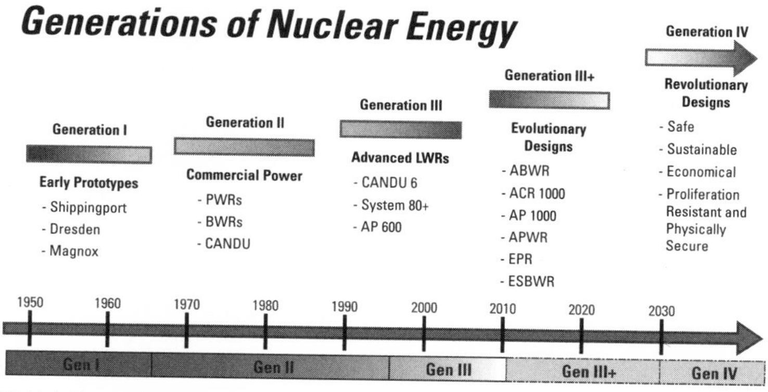

FIGURE 14-1: The evolution of nuclear reactors according to the Generation IV International Forum.

Generation IV consists of advanced reactor types that are being studied by the GIF. At this time, a group of improved designs evolved from Generation III reactors, called Generation III$^+$, are being designed and constructed throughout the world.

The new reactors to be built in Canada will likely be of the Generation III$^+$ type. In particular, we will concentrate our discussion on the three systems in the running for the new Ontario reactors.

The Advanced CANDU Reactor (ACR-1000)

The Advanced CANDU Reactor (ACR) is in the conceptual and research phase with the design scheduled to be completed by 2012. It will be heavy-water moderated and light-water cooled. One objective is to significantly reduce the volume (and therefore the cost) of heavy water. Its design power is 1,050 megawatts.

A variety of possibilities is being considered to reduce costs, enhance the engineering process, and include more passive safety systems. For example, a new fuel bundle called CANFLEX® has been developed with 43 fuel elements, instead of the 37 currently used. This has been successfully tested at the Point Lepreau reactor. In addition, the elements have two different diameters as well as special appendages to improve the transfer of heat from the bundle into the reactor coolant. Unlike current CANDUs, the ACR will use slightly enriched uranium fuel (around 2 percent uranium-235) so that more electrical power can be generated in each fuel channel, leading to a smaller reactor core. In turn, many of the other systems can also be scaled down. A smaller and simpler reactor means maintenance and reliability will improve, while cost will be reduced.

As an example of the value of a smaller reactor, a CANDU-6 reactor requires 460 tonnes of heavy water (265 tonnes as moderator and 192 tonnes as coolant). The ACR-1000 as currently envisaged will require only 250 tonnes of heavy water, even though it has significantly higher output power (1,050 megawatts versus an average of 640 megawatts for the CANDU 6).

Furthermore, new methods of producing heavy water at less cost and avoiding the environmental problems associated with the hydrogen sulphide used in former heavy-water plants are now under development. A prototype heavy-water plant in Hamilton, Ontario, used the Combined Industrial Reforming and Catalytic Exchange process. With a production of 1 tonne of heavy water per year, this plant operated successfully for about two years ending in 2002. In 2008 there were no heavy-water production plants in Canada.

Due to its good neutron efficiency, potential for using advanced fuel cycles, and strong high-technology research support, many improvements are possible to the CANDU reactor that will increase its safety, efficiency, and reliability, while reducing both its capital and operating costs. However, like its CANDU predecessors, the ACR-1000 will require replacement of its pressure tubes after thirty years.

The ACR-1000 is about four years behind its Ontario competitors, the Westinghouse AP1000 and the AREVA EPR reactors, which in 2008 were already being built.

The Westinghouse AP1000

In early 2008, construction began on four AP1000 reactors by two utilities in China. In addition, in 2008, there were two orders, each for two AP1000 reactors, by utilities in the United States. The AP1000 is well on its way to becoming a reality.

The AP1000 is an advanced pressurized-water reactor (PWR). It uses enriched uranium fuel and is both cooled and moderated by water. It also has passive safety features. Passive plants are dramatically different from the current generation of reactors. They have been designed so that their safety systems respond to accidents automatically, without human or motor intervention. They do this using the natural forces of physics, gravity, and convection; no electricity is required except for what has been stored in emergency backup batteries. The reactor will automatically shut down in the event of an accident, and the decay heat from the fuel will be automatically dissipated without damaging the fuel or reactor core. Fuel melting and, consequently, major accidents become much less likely. These designs are quite remarkable and should go a long way toward calming the public's fear of catastrophic reactor accidents.

The main method for heat dissipation is natural convection: as fluids become warmer they expand, become less dense, and rise.

Similarly, as fluids become colder they contract, become denser, and sink. Convection is responsible for chimneys sending smoke and hot air up and out of houses.

A key feature of the design is the ability to remove heat in the case of an accident. The AP1000 does this in an ingenious manner. In an accident, heat from the reactor fuel elements would cause some of the water covering the core to evaporate. The steam that forms would condense on the walls of the steel containment, but only if the walls are cooled. So it is essential to ensure that the walls are kept cool. The reactor containment consists of a steel shell located inside a concrete building. The building has air inlets as well as a baffle between the building and steel containment vessel that is open at the bottom. Air flows, by natural convection, downward between the concrete building and baffle, makes a U-turn at the bottom. It then flows upward between the baffle and steel containment to exit at the top of the building. Air is heated as it passes the hot containment shell and rises like smoke in a chimney. At the same time, cold air is drawn from the outdoors to form a natural convection loop without the need for pumps or fans (see Figure 14-2).

To augment the removal of heat, a large tank of water is situated on top of the building where valves automatically sense a rise in temperature and pressure and release water onto the steel containment shell for the first few days after an accident. The water is vaporized, drawing off additional heat, and the steam flows out of the top of the building. Note that neither the air nor the water that is used for cooling the steel containment shell comes in contact with radioactive materials.

The preceding discussion illustrates only a few of the new and innovative features in the next generation of nuclear passive reactors.

The AREVA EPR

The AREVA EPR ("Evolutionary Pressurized Reactor") is also a Pressurized Water Reactor (PWR). In comparison to the AP1000, the EPR relies more on engineered safety features than on passive safety systems.

The EPR evolved from the last four French reactors that were completed in 2000 as designed by Framatome (now part of AREVA). The reactor uses regular (or light) water for both moderation and

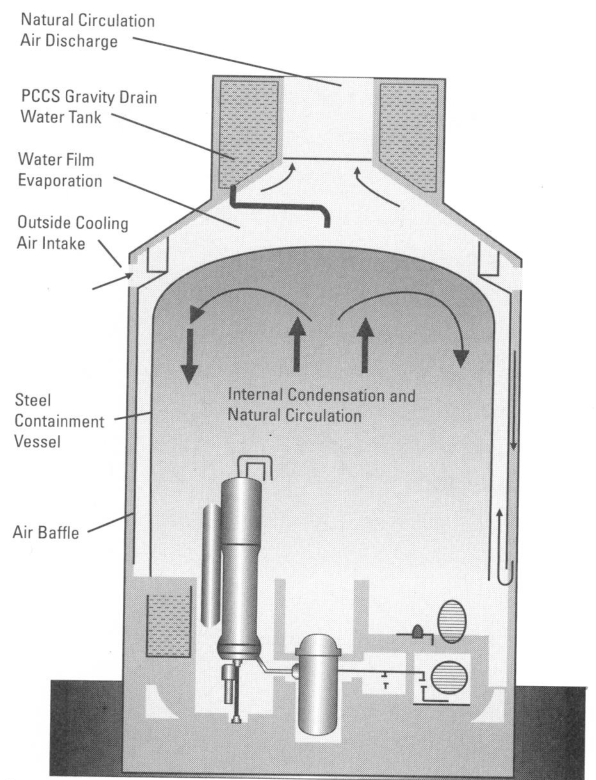

FIGURE 14-2: Schematic of the AP1000 reactor.

coolant. Its fuel is enriched uranium with up to 5 percent uranium-235. In these respects, it is generically similar to its Westinghouse rival. However, its power output of 1,600 megawatt is much greater than that of the AP1000 (about 1,100 megawatt) and the AECL ACR-1000 (about 1,050 megawatt).

The EPR has a complex containment system consisting of a steel shell attached to a concrete shell that is partly designed to guard against potential aircraft strikes. To counter an accident in which a hot reactor core of molten fuel might burrow into the earth—the notorious "China Syndrome"—the EPR has a "core catcher" consisting of a specially designed concrete basin. Other features include separate compartments for the heat transport (coolant) pumps and a pool of water at the base of the reactor. It also has a pool of cooling water in the roof of the reactor building, a feature it shares with the AP1000.

The EPR is the farthest along of the three reactor types under consideration for Ontario. Construction of an EPR at Olkiluoto, Finland, commenced in 2005. This was the first new reactor ordered in western Europe in the last decade. Originally scheduled to begin commercial production in 2009, it was two years behind schedule by 2008 with a 50 percent cost overrun. Startup is now estimated in 2011. A second EPR is also under construction at Flamanville, France.

In addition to the three reactor types described above, it should be noted that there are other Generation III$^+$ designs: for example boiling water reactors such as the GE-Hitachi Economic Simplified Boiling Water Reactor.

Generation IV Reactors

The purpose of the GIF is to study future reactors that are safe, secure, and economic. Some will use recycled fuels to be environmental sustainable but in such a way as to be resistant to weapons proliferation. The advanced reactor types being considered are:

- Very-High-Temperature Reactor: a graphite-moderated, helium-cooled reactor with no fuel recycling.
- Supercritical-Water-Cooled Reactor: a high-temperature, high-pressure water-cooled reactor. The high temperatures and pressures yield greater efficiency.
- Gas-Cooled Fast Reactor: a fast reactor, helium-cooled with recycled fuel.
- Lead-Cooled Fast Reactor: a fast reactor with lead (45 percent)/bismuth (55 percent) liquid metal coolant and recycled fuel.
- Sodium-Cooled Fast Reactor: a fast reactor, sodium-cooled reactor with fuel recycling.
- Molten Salt Reactor: produces fission power in a circulating molten salt fuel mixture which is also the coolant with fuel recycling.

The partner countries in GIF generally do research relevant to only a few of the above six types. Canada is specializing on the Supercritical-Water-Cooled Reactor.

Most of the Generation IV reactors are of the type known as fast breeders. A few examples have been in operation for many years. Today's fast breeders have no moderator, use enriched uranium or plutonium fuel, and use liquid sodium metal as coolant. The descriptor "fast" refers to the high-energy, or fast, neutrons that arise because there is no moderator to slow them down. These reactors create more fuel as they operate, converting uranium-238 to plutonium-239, hence the name "breeder." The central part of the core is enriched to 10 to 20 percent fissile material (either uranium-235 or plutonium-239). Surrounding this is a "breeding blanket" of natural uranium. During operation the uranium-238 in the blanket is irradiated, absorbs neutrons, and transforms to plutonium-239.

This reactor has promise because not only is uranium-238 abundant and generally of little value (huge stockpiles of depleted uranium, i.e. with the uranium-235 largely removed, have been built up at uranium enrichment plants around the world), but it is also possible to make more fuel than is used. A virtually inexhaustible supply of nuclear fuel would be available.

This type of reactor must be cooled by liquid sodium, a material with low absorption and low moderation of fast neutrons. A drawback to the fast breeder is that sodium leaks have caused problems in several prototype reactors. The most recent was in 1995 at Japan's Monju fast breeder reactor, which suffered a serious sodium leak and fire after one year of operation. Work is in progress to re-start the reactor.

Russia, France, Japan, and India now operate breeder reactors. Canada has not, although the CANDU can be converted into a near-breeder reactor, converting thorium-232 fuel to uranium-233. There is now more interest in this technology in view of the nuclear renaissance that has come about with the selection of Generation IV designs.

-15-

Fusion:
The Energy of the Future?

Fusion, the energy that powers the sun and the stars, is the ultimate energy of the universe. In fact, most of our present forms of energy are fusion based, of which solar energy is the most obvious example. Fossil fuels (coal, oil, and natural gas) all originated from ancient plants that grew in sunlight. The sun's energy causes the rains that fill the reservoirs of hydroelectric dams and the winds that turn the blades of wind turbines. All living things on earth owe their existence to fusion.

Humans have long had the dream of controlling fusion. With the oceans holding an endless reservoir of hydrogen—the fuel for fusion—we also have the promise of an unlimited supply of cheap fuel. Although progress has been made, the road to harnessing fusion power has been rocky and frustrating. Humans so far have been unable to control a fusion reaction (although uncontrolled fusion has been used in the hydrogen bomb).

Fusion Basics

Fusion, like fission, is a nuclear process involving interactions between the nuclei of different atoms. Instead of splitting a heavy nucleus, two light nuclei are joined together to form a single one. For some of the light elements, the mass of two individual nuclei is greater than the mass of the combined nucleus. In a fusion reaction, the surplus in mass is converted to energy, according to Einstein's equation. The fraction of matter turned into energy is greater than in fission, making it a process with far more potential for energy generation.

All the light elements up to the mass of iron (atomic number 26) are theoretically capable of undergoing fusion and producing energy. In certain stars, all of these fusion reactions take place. Under conditions we might produce on earth, however, only the three lightest elements, hydrogen, helium, and lithium, have a chance of fusing.

The nuclei of light elements do not join together easily. To convince them to do so, enormous temperatures are required. The fusion reaction that occurs at the lowest temperature, and thus is of most interest, is the deuterium-tritium reaction:

$$^{2}\mathrm{H} + {}^{3}\mathrm{H} \longrightarrow {}^{4}\mathrm{He} + \text{neutron} + \text{Energy}$$

The same reaction is also written:

deuterium + tritium ⟶ alpha + neutron + Energy

In this reaction, deuterium and tritium represent the two isotopes of hydrogen, respectively; and alpha represents an alpha particle, or helium nucleus.

Deuterium is readily obtained from any source of water: some 150 out of a million hydrogen atoms in nature are deuterium atoms. As a large fusion reactor (1,000 megawatts) would only use a few hundred kilograms of deuterium per year, a single heavy-water plant would supply many "deuterium-tritium" fusion reactors.

It is proposed that tritium be made by placing a "blanket" of natural lithium (with three protons) in the fusion reactor where some lithium atoms will absorb neutrons and undergo a nuclear reaction to form helium and tritium. The tritium will be separated from the lithium by chemical processing. Lithium is a common element in the earth's crust, much more common than uranium. Only a relatively small amount of lithium would be consumed by a fusion power reactor and for this reason, there are essentially no resource limits on fusion fuel for far into the future—clearly, a major attraction of fusion.

The deuterium-tritium reaction requires a temperature of about 80 million °C. This temperature would vaporize any kind of a container, preventing any kind of extraction of energy to drive turbines and generate electricity. So how can a fusion reaction be contained?

The answer is to create a "plasma," which can be contained by magnetic fields rather than physical walls. By raising the temperature of a gas high enough, the atoms will collide with sufficient energy to remove their electrons. The resulting high temperature medium, plasma, consists of positively charged ions and negatively charged electrons, whose overall electrical charge is neutral. Plasma has quite different properties from the gas out of which it was formed. For example, plasma is a very good conductor of electricity because of its free electrons and charged nuclei, whereas the gas it was made from may be a very poor conductor. It is also a good heat conductor, while gases are generally poor in this respect.

Plasma has often been called the fourth state of matter. It exists on earth in such diverse places as welding arcs, lightning discharges, and inside fluorescent-light fixtures. Stars are made up largely of matter in the plasma state. Charged particles from outer space create plasmas by travelling into the earth's atmosphere along its magnetic field lines. The light radiated by these plasmas is called the Northern Lights (see Chapter 3).

Confinement of plasmas at the necessary enormous temperatures takes place in two ways. One is to confine the plasma in a web of magnetic fields, known as Magnetic Confinement Fusion (MCF). The other approach to fusion is called Inertial Confinement Fusion (ICF). In this method, a great deal of energy is used to create plasma in a very short time and the fusion process is completed so rapidly that the plasma particles are essentially confined by their own inertia.

Inertial Confinement Fusion

Although most of the international fusion effort is concentrated on magnetic confinement, it is helpful to an overall understanding of fusion to briefly describe inertial confinement.

In inertial confinement fusion, a small pellet consisting of deuterium and tritium is rapidly heated and compressed to a high density by a focused implosion of the pellet. The objective is to raise the temperature and density high enough so that fusion will be achieved for the short confinement time determined by the inertia of the pellet.

In one method, many intense laser beams are directed simultaneously onto a small spherical fuel pellet from all sides. The pellet may simply be a small shell of glass or plastic filled with a deuterium-tritium mixture perhaps a few millimetres in diameter or an elaborate multi-layered structure designed to promote the implosion and minimize losses. Figure 15-1 shows a demonstration of fusion in such a pellet.

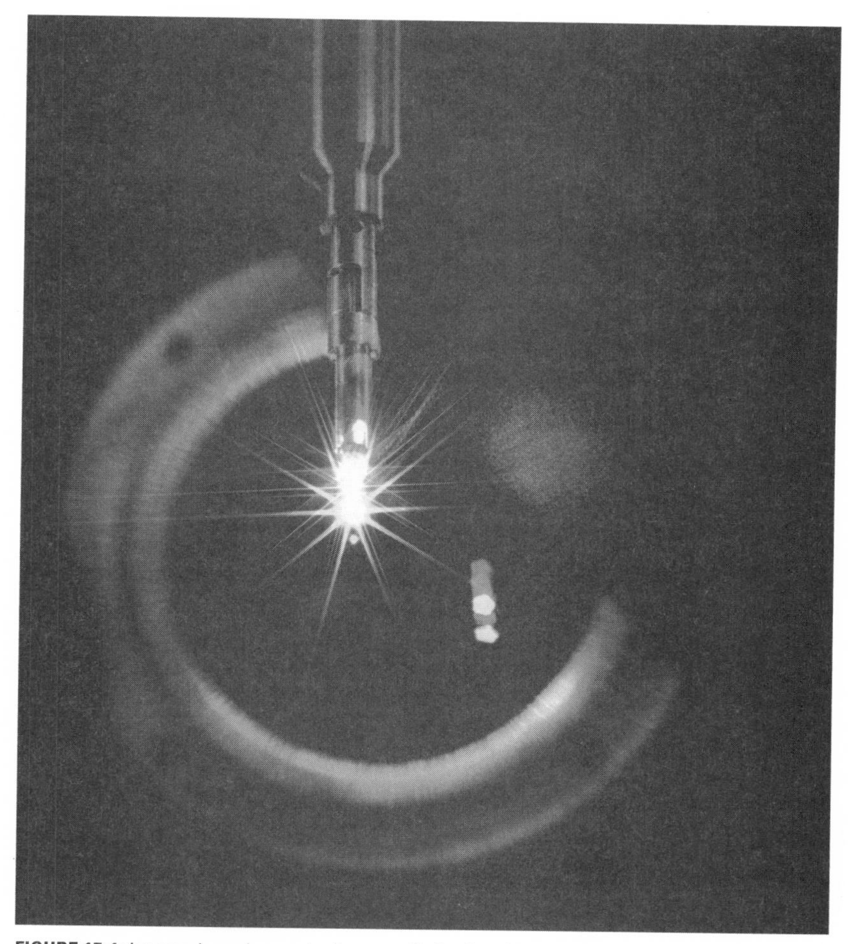

FIGURE 15-1: Intense laser beams ignite a small plastic sphere containing deuterium and tritium producing a fusion reaction.

The initial intense irradiation of the fuel pellet vaporizes material from the pellet's surface and turns it instantaneously into a corona of plasma blasting away from the still solid pellet. The resulting shock wave compresses the fuel in the pellet to a high density (a thousand or more times liquid density). Laser energy must continue to be absorbed by, or coupled to, the outer corona of plasma in order to continue the compression of the pellet.

One of the basic problems is to achieve efficient coupling of the energy of the lasers to the pellet. It is necessary that the energy be largely used to produce the implosion rather than being dissipated by various non-productive mechanisms. Another important challenge is to construct laser or particle beams of sufficient power, uniformity, energy efficiency, and repetition rate to allow a practical fusion system.

Inertial confinement fusion is perhaps a decade or two behind magnetic confinement fusion in the progress achieved to date. However, the potential for military applications continue to ensure a high level of interest in high-energy lasers. They are used to simulate various aspects of nuclear weapons now that atmospheric weapons testing has been banned. The first major inertial confinement installation, called Shiva, was constructed at Lawrence Livermore National Laboratory in California and uses twenty separate laser beams. An even larger installation called Nova, which is twenty times more powerful, has been in operation there for over a decade. Other major ICF facilities are Omega at the University of Rochester and Gekko XII in Japan. Much larger laser ICF installations, the National Ignition Facility in the United States and the Laser Mega Joule project in France, costing in excess of US$3 billion each, are now under construction. The National Ignition Facility was essentially complete in 2008 and full ignition of a deuterium-tritium pellet is expected by 2009–2010.

Magnetic Confinement Fusion

Magnetic confinement fusion (MCF) has been pursued continuously since the 1950s and is the method on which most of the world effort is concentrated. MCF depends on the fact that the electrically charged plasma particles (positively charged hydrogen ions and negatively charged electrons) gyrate around magnetic field lines in a corkscrew fashion. In this way, the plasma can be confined and kept away from the container walls by applying suitably shaped magnetic fields. During this so-called confinement time, the temperature and density of the plasma must be increased to the point where fusion can take place.

Confining a plasma using magnetic fields is very difficult and has been said to be "like confining Jell-O with elastic bands." Decades of research have been required to understand the origins of various instabilities and how to overcome them. There are many possible configurations of MCF devices depending on what arrangements of magnetic fields are used. Early researchers contained plasma using cylinders.

These were wrapped with electrical conductors through which electricity was passed to create a magnetic field in the cylinder. These open or linear systems initially had great appeal due to their relative simplicity. However, techniques were never found for preventing a significant amount of plasma from leaking out the ends of the cylinder.

This problem was solved by making closed, toroidal (doughnut-shaped) systems. Magnetic coils surround the torus chamber containing the plasma and they create a uniform magnetic field along the axis of the torus. This toroidal field alone, however, is not sufficient to contain the plasma, as it will drift toward the walls of the torus. The addition of another magnetic field, called the poloidal field, in the direction of the small circumference of the torus is required. The resultant spiral field, obtained by adding the two fields, effectively contains the plasma.

Tokamaks

To date, the best device for confining plasma is the tokamak, which was invented in Russia in the 1960s. Tokamak is an abbreviation for a Russian phrase meaning "toroidal magnetic chamber."

The toroidal field coils are super-conducting magnets arranged so that a toroidal or doughnut-shaped volume is created. The plasma current circulates like a merry-go-round in this volume and becomes the secondary circuit of a transformer. The primary circuit of this transformer is formed from other coils, called poloidal, placed around the doughnut. The resulting current heats the plasma, the so-called ohmic heating, but this method alone cannot easily be used to attain fusion temperatures. The magnetic field from the current itself and the primary circuit coils helps confine and shape the plasma, and to keep it positioned in the toroidal volume. The total magnetic field from the toroidal and poloidal coils and from the plasma itself describes a toroidal helix.

Because of its reliance on transformer action to create the necessary fields, the basic tokamak design is limited to pulsed operation. This means that they create, confine, and heat a plasma for at most a few tens of seconds, at which point the plasma leaks away and the machine must the be prepared for another pulse. Obviously this is an

FIGURE 15-2: Exterior and interior views of the JET tokamak, illustrating the size of this device. The volume of the plasma in the ITER machine will be about 10 times larger.

undesirable manner in which to operate an electrical power station. Various non-inductive means for sustaining the plasma current are being explored to see whether steady state, or at least long-pulse operation, is feasible.

A number of tokamaks have been built including the Joint European Torus (JET) in England which is operated by the European Community. The scale of the JET machine can be appreciated from Figure 15-2. The Japan Tokamak in Naka, Japan and the American Tokamak Fusion Test Reactor (TFTR) at Princeton University, New Jersey, which operated from 1982 to 1997, are the other two large tokamaks that have been built. Two other major tokamaks are the DIII-D at General Atomics in San Diego and the Tore-Supra at Cadarache, France. The Canadian research tokamak, the Tokamak de Varennes near Montreal, operated from 1988 until 1997 when it was shut down and dismantled.

Energy breakeven, a measure of fusion performance, means the fusion energy produced is exactly equal to the energy used to heat the plasma to get the fusion reaction to occur. All three large tokamaks have created plasma conditions with deuterium plasmas that would give energy breakeven if tritium had been used in a deuterium-tritium reaction. The current record is held by the Japan Tokamak, achieving breakeven plus 25 percent additional energy. There is a reluctance to use much tritium in JET and TFTR because these machines were not designed to deal with any but a nominal radiological hazard. It is forbidden by Japanese Law for the Japan Tokamak to use any tritium at all. Nevertheless, both JET and TFTR ran a few experiments with deuterium-tritium mixtures, which produced a few tens of megawatts of power.

Other Approaches to Magnetic Confinement Fusion

In addition to the problems caused by pulsed operation, tokamaks have generally not been very efficient at keeping energy in the plasma. For this reason research on other types of toroidal devices is being pursued throughout the world.

The most developed MCF systems after the tokamaks are stellerators, which use relatively complicated systems of conductors arranged around the torus to create the twisted magnetic fields needed for stable confinement. Since they do not use the transformer effect, stellerators should run continuously instead of being pulsed. A large stellerator called Large Helical Device was completed in Japan in 1998, and a stellerator of another type called Wendelstein VII-X is under construction in Germany. These devices represent investments of about US$1 billion each. At this time, stellerators are performing roughly a factor of 10 behind tokamaks in terms of energy breakeven.

Several other types of MCF systems are being pursued at relatively low levels of funding. As yet, however, they are very far from the level of performance of today's tokamaks.

Canada's Role in Fusion Research

Canada was an active and internationally recognized contributor to fusion research from the early 1970s until 1997, although our contribution was small compared to that of the major players.

The Canadian Fusion Fuels Technology Project was located in Mississauga, Ontario and conducted research based on the unique Canadian expertise in the primary fusion fuels, deuterium and tritium. The studies focused on tritium and fusion fuel systems including tritium blanket studies. Although all of Canada's heavy-water plants are now closed, a large quantity of heavy water has been stockpiled, so that deuterium is readily available. Tritium is produced as a by-product in CANDU heavy water. With 18 CANDU reactors, Canada is the largest non-military producer of tritium. Furthermore, technology has been extensively developed for the safe handling of tritium in terms of worker safety and environmental impact.

Special magnetic confinement and materials studies were conducted at the Tokamak de Varennes from 1988 to 1997 (Figure 15-3).

Both Canadian fusion projects were closed when federal government funding was terminated in April 1997, due to deficit reduction measures. A small amount of fusion research is still done in Canadian universities, primarily at the Universities of Saskatchewan, Toronto, and Quebec.

The Prospects for Fusion Energy

A practical fusion reactor will need more than just the success of the plasma part of the system. A formidable challenge will be to collect the energy and convert it to electricity.

FIGURE 15-3: The Tokamak de Varennes was Canada's state-of-the-art fusion device located near Montreal. Built at a cost of $100 million, it showed excellent performance until it was closed in 1997 due to government budget cuts.

The conversion of plasma energy to electricity will take place via the lithium blanket, which can be in solid or liquid-metal form. The neutrons from the plasma will interact with and heat the lithium whose coolant (liquid lithium, helium, or water) will be pumped through a heat exchanger where it will transfer the heat to steam for driving turbines. There will be major materials problems. New alloys are needed that can withstand bombardment by high-energy neutrons from the fusion reaction. Another problem concerns the erosion of the plasma-chamber walls by particles that escape the confinement of the magnetic fields. Wall design is a major challenge, and it will be necessary to replace it periodically during the life of the plant. Of course, the whole reactor will become radioactive because of nuclear reactions initiated by the fusion neutrons and so remote manipulation will be needed to service it.

A fusion reactor will generate radioactive wastes, but these will be smaller in quantity and less radiotoxic than those generated by current fission reactors of similar power. Furthermore, tritium, which is a radiological hazard and requires careful control, will also be involved.

In 1985, the International Thermonuclear Experimental Reactor (ITER) project was initiated jointly by the United States, the European Community, Russia, and Japan. Canada participated through the European Community until 1997. The objective of ITER is to demonstrate the scientific and technological feasibility of fusion power and to provide the technical database for the design and construction of a demonstration fusion power plant. ITER is an enormous undertaking, certainly the largest energy research project in history, costing over

Cold Fusion

On March 23, 1989, two chemists from the University of Utah held a press conference at which they stunned the world by announcing that they had created fusion in a test tube. This experiment, quickly labelled "cold fusion" by the media, became the science story of the decade and set off frantic efforts in other world laboratories to duplicate it. The stakes were very high since this phenomenon was touted by its proponents as the solution to all the world's energy problems.

In the weeks following the announcement, hundreds of laboratories tried the experiment but only a very few claimed to observe any fusion effect. Some chemists tried to become instant experts in fusion physics while some physicists made elementary blunders in chemistry with many scientific reputations damaged by sloppy experimentation and unfounded theorizing. Eventually, it became apparent that cold fusion was an illusion seen only by a small number of enthusiasts. Cold fusion is best remembered as an example of bad science, and a warning to all scientists not to let their enthusiasm blind them to physical realities.

US$12 billion for the project lifetime of thirty years. After a lengthy process of design and negotiations a treaty was signed in 2007 between the original four parties plus China, India, and Korea to implement the ITER project at Cadarache in the south of France. Preparation of the site is underway and long lead-time components are being ordered. Even if ITER is successful, another step would be needed: a demonstration power reactor to ensure that the new materials developed for fusion would actually survive in a working reactor.

The scientific feasibility of fusion energy has been demonstrated in the current generation of large tokamaks. Nevertheless, a great deal of engineering development, lasting for at least 50 years, will be needed for an economic and reliable fusion power system. For this reason, we should not expect to see fusion producing any substantial percentage of the world's energy before the end of this century.

-16-

Research: The Path Forward

Canadians are more than hewers of wood and drawers of water. Ingenuity and determination have opened up the country through such engineering marvels as the trans-Canada railway that stretches from coast to coast and the St. Lawrence Seaway that allows giant ocean-going ships to sail into the centre of North America. Scientists, engineers, and entrepreneurs have converted our resources into complex, high-technology products. And the more complex and ingenious the products become, the more value they gain.

It is the combination of vast resources and creative minds that have brought prosperity to Canada. Our nation is among the world leaders in many high-technology fields such as telecommunications and robotics. Several Canadians have been awarded Nobel Prizes in the sciences.

The Canadian nuclear industry is an excellent example of how research and technical innovation yield new products and economic growth. Nuclear research and development resulted in the unique CANDU reactor that supplies electricity in Canada and in four other countries. Its spinoffs include radioisotopes and cancer therapy machines that have transformed the way that medicine is practised. Hundreds of ingenious applications of nuclear gauges and radioisotopes are used in the industrial sector; these all improve the products we use. And there have been many other advances such as simulator models, computerized control systems, and fundamental progress in metallurgy and chemistry.

The National Nuclear Laboratory

It is only when enough nuclear fuel is assembled—an amount that equals or exceeds the critical mass—that a chain reaction will occur. In an analogous manner, an intellectual critical mass can be created by assembling a large group of top-class scientists and engineers,

FIGURE 16-1: Chalk River Laboratories is located on the banks of the Ottawa River about 200 kilometres northwest of Ottawa. The two largest buildings in the centre house the NRX and NRU reactors.

equipping them with the best facilities available, and concentrating their efforts on a particular area of science and technology. This concept for excellence in research and development is known as a national laboratory and has been used to advance important technology sectors in many countries.

In Canada, there has been a national nuclear laboratory since 1944: the Chalk River Laboratories at Chalk River, Ontario (shown in Figure 16-1). The Chalk River laboratory is a large facility (2,000 people) with a rich history. Chalk River made fundamental contributions to the development of the CANDU reactor starting with the creation of ZEEP in the 1940s and continuing with the NRX and NRU reactors. Another research facility, Whiteshell Laboratories, also part of AECL, is located at Pinawa, Manitoba. Nuclear research is also carried out at universities and hospitals.

AECL is one of only two government agencies operating major research laboratories in Canada (the National Research Council is the other). The contribution that these laboratories have made to the Canadian economy through basic and applied research is difficult to measure, but is certainly significant.

The discovery of channelling is an excellent example of the interplay between science and applied technology seen in a national laboratory. Nuclear physicists needed to know how far into a solid an ion could penetrate. Determining this quantity, known as "the range," by theory was very difficult since it involves a variety of complex interactions. Experiments were necessary.

In the early 1960s, chemists in Chalk River used techniques that allowed them to measure very small amounts of radioactive materials dissolved in liquids to measure the ranges of ions in solids. First they implanted radioactive ions in a solid using an ion accelerator. Then they dissolved a very thin layer of the solid. The radioactivity of the dissolved solid layer was measured and the number of radioactive ions in that layer was determined. This process was repeated, peeling off successive layers of the solid. In this way a range profile showing the number of ions that penetrated to each depth was built up. By repeating the experiment for many different ion-solid combinations, the main features of ranges gradually emerged.

The results of most of these painstaking experiments could be explained by existing theories and it looked as if the subject was well understood. In a small percentage of the experiments, however, the ions were penetrating much deeper than predicted. As often happens in science, this small anomaly turned out to be extremely important.

At about the same time, scientists at Oak Ridge National Laboratory, one of the national nuclear laboratories in the United States, began computer simulations of this phenomenon using the large high-speed digital computers that had just become available. In today's terms, they set up a virtual solid consisting of an arrangement of atoms and then moved virtual ions through it using Newton's laws of motion to solve the collisions between the ion and the solid atoms. They did this for many ion-solid combinations, effectively performing computer experiments equivalent to the physical experiments going at Chalk River. The Oak Ridge scientists also observed the anomalously long ranges in a few cases. When the two groups compared results they confirmed that they were observing a new physical phenomenon, not an experimental error.

Figure 16-2 depicts an ion travelling down a "channel," or tunnel, in a solid material. Crystalline solids have regular arrangements of atoms in a three-dimensional pattern (called a lattice) and if an ion is lined up just right with respect to such a tunnel, it can travel long distances into the solid.

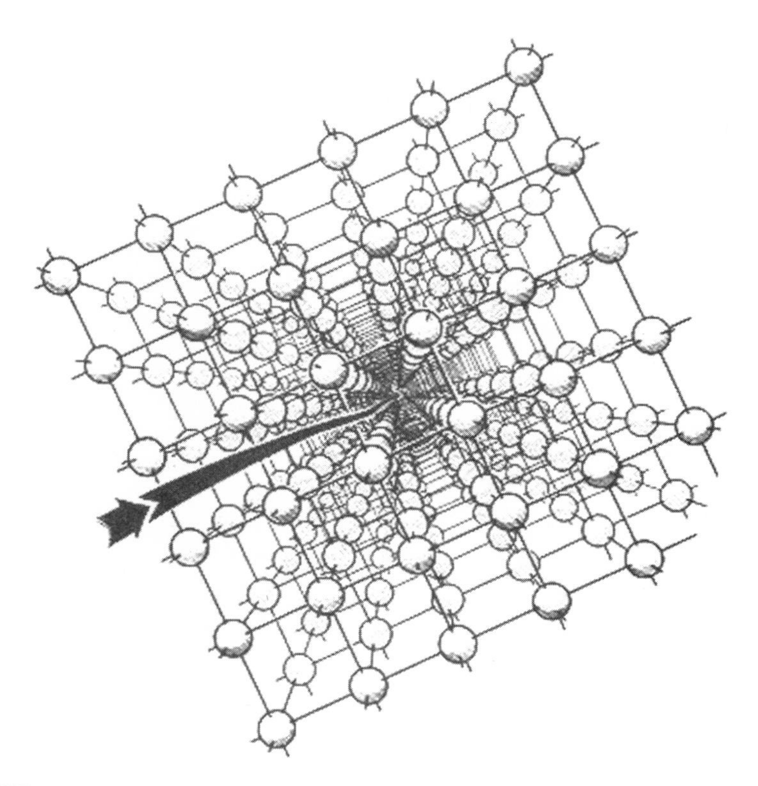

FIGURE 16-2: Illustration of channelling showing an ion being steered down a solid lattice.

At first channelling was thought to be just another scientific curiosity. It was soon realized, however, that channelling had very important applications. The use of ion implantation to inject trace materials ("dopants") into semiconductors was just beginning, and the discovery of channelling initiated a scientific and technological "boom." All over the world, scientists from universities and research institutes as well as from semiconductor, telecommunications, and computer companies were fascinated. Many came to visit Chalk River and set up collaborative research programs. A new area of science and technology involving hundreds of researchers globally was launched and a series of international conferences was initiated. Applications of ion implantation are not just confined to semiconductors, but also impact areas such as metal hardening for wear resistance (surface alloying), plasma processing, fusion, and nuclear physics. For example, hundreds of specialized ion implanting accelerators are now used throughout the world to manufacture integrated circuits. The basic nuclear research at Chalk River made a significant contribution to the computer/communication revolution.

This Chalk River story is a good illustration of scientific research—the way that one area of research leads to another, its international aspect, the need for the scientific freedom to pursue the unexpected, and the close relationship between science and technology. It also shows the value of scientists skilled in different disciplines working together in an open intellectual environment and with sound financial support—the ideal situation for a national laboratory.

Nuclear Science

Scientists love firing particles at atoms and studying what happens. A favourite particle is the neutron that, lacking an electrical charge, can pass by the electrons to interact with the nucleus. When a neutron is captured by a nucleus it often upsets the equilibrium: the neutron is converted into a proton. To conserve electrical charge, a beta particle is also created and is ejected from the nucleus. The proton stays in the nucleus, creating an element with an atomic number one higher than before. For example, a hydrogen atom that gains a proton by this process becomes a helium atom. A helium atom would become a lithium atom and so on.

Uranium is the largest atom found in nature; no natural element contains more protons. Is it possible to create a brand new element heavier than uranium? That is exactly what happened in the 1930s when scientists bombarded uranium with neutrons. They formed an element that no one had ever seen before, with 93 protons, and they named it neptunium, since Neptune is the planet beyond Uranus. And when neptunium decays by beta emission, it transforms to a new element with 94 protons called plutonium (Pluto is the planet beyond Neptune). Both neptunium and plutonium are radioactive.

These fascinating developments in nuclear physics were also explored at Chalk River. The first phase of the Tandem Accelerator and Super Conducting Cyclotron (TASCC) was completed in 1987, and full operation began in 1991. It consisted of an accelerator, which accelerated heavy atoms and then fed them into a cyclotron where they were further accelerated to almost the speed of light. The cyclotron used magnets made of super-conducting materials, which conduct electricity thousands of times more efficiently than ordinary wires. The building of the superconducting cyclotron, the largest such facility in Canada, was a major technological feat in its time. The accelerated heavy atoms were fired at target nuclei and the interactions were studied. The accelerator/cyclotron made many original contributions to nuclear physics before it was closed in 1997 due to federal budget cuts.

Another impact of basic nuclear research facilities at Chalk River was the training of researchers who subsequently moved to the physics and chemistry departments of many Canadian universities. This was particularly pronounced during the large expansion of Canadian universities in the 1960s. Even today, most Canadian physicists have had some sort of past association with Chalk River or one of its offshoots.

Physicists from Chalk River, for example, sparked the building of the "Tri University Meson Facility" (TRIUMF), which started in 1975. This facility is a collaboration between the University of British Columbia, the University of Victoria, Simon Fraser University, and the University of Alberta; as such, TRIUMF is the leading centre for particle physics research in Canada. This machine uses electromagnets to force positively charged hydrogen atoms into circular paths at speeds up to 200,000 kilometres per second (67 percent of the speed of light). These ion beams are then shot onto special targets to study the subatomic fragments that result from the collision. Exotic subnuclear particles, such as mesons, pions, muons, and neutrinos, are formed which only theoretical physicists can love or understand. TRIUMF also produces radioisotopes that are deficient in neutrons such as cobalt-57, gallium-67, and indium-111. These isotopes are packaged and marketed by MDS Nordion.

Bertram Brockhouse (1918–2003) was born in Alberta and educated at the Universities of British Columbia and Toronto. From 1950 to 1962, he made ground-breaking discoveries in the physics of solids using neutrons from Chalk River's NRX and NRU reactors, in the process developing a new instrument for neutron scattering, the triple-axis spectrometer, which for the first time permitted the measurement of the changes in energy of neutrons after they have moved through matter. He was the first to measure the energy-versus-momentum (or the frequency-wavelength) relationship for lattice waves in crystals, liquids, and magnetic materials. He received many honours including Fellowship in the Royal Society of London, the Order of Canada, and the joint award of the Nobel Prize in physics in 1994. Until his death he was Professor of Physics at McMaster University, which he joined in 1962.

The Sudbury Neutrino Observatory, another basic-science facility involving graduates of Chalk River, began operating in 1999. Built deep in the Creighton nickel mine in Sudbury, it measures the neutrinos emitted by the sun and other stars. The number and the type of neutrinos indicate the nature of the nuclear reactions taking place in stars, and give scientists insights on the evolution and composition of the universe. Neutrinos are very difficult to observe since they tend to pass through material objects. The observatory is located deep in a former nickel mine where energetic cosmic-ray particles cannot penetrate and interfere with the experiment. The detector is an acrylic plastic sphere filled with 900 tonnes of heavy water, and that is surrounded by ordinary water. Highly sensitive light detectors surround this sphere in a totally dark cavern and detect the photons emitted in the rare events when a neutrino hits one of the nuclei in the heavy water.

In June 2001, the SNO team announced that their observations confirmed that the neutrino had a mass, a result first reported in 1998 from Super-K (a similar Japanese experiment). They also observed all of the types of neutrinos emitted by the sun, something not possible at Super-K. From this, they concluded that the sun's output of neutrinos was in excellent agreement with theoretical predictions. These announcements were made to international acclaim because of their profound significance to fundamental physics, as well as our understanding of the universe. The SNO was closed in 2007.

Neutrons, according to quantum physics theory, have the properties of both a particle and a wave. For thermal neutrons (i.e. with low energy), the wavelength happens to be similar to the spacing of atoms in a crystal. This allows them to be diffracted by a crystal in the same way that light is diffracted by a prism or lines on a grating. Neutrons, then, can be used to study the structure of atoms. The neutron scattering program at Chalk River is now operated as a component of the National Research Council of Canada.

Innovation from Nuclear Research

Previous chapters have described the nuclear technology's impact on energy, medicine, industry, and science. Here we will describe some specific examples of the innovations that have flowed from the Canadian nuclear research and development program.

Research into the effects of radiation on human health has been conducted at Chalk River for over 35 years. It was at Chalk River that the ideas for cancer therapy using cobalt-60, positron emission tomography, and the thermoluminescent badge for personal dose monitoring originated. Research into molecular biology and mechanisms of cancer induction have made significant contributions in the campaign to find a cure for cancer. In 2000, these research areas were transferred from AECL to the federal Department of Health.

From its outset, Chalk River has conducted research into the use of radionuclides for industry and medicine, as well as producing radionuclides in its reactors. Some years ago Nordion was spun out of AECL as an independent business. It was subsequently brought by MDS, renamed MDS Nordion, and has become an international leader in the medical and industrial application of radiation.

In the 1990s, MDS Nordion funded AECL to design and build the world's first dedicated reactors for producing radioisotopes. However, these MAPLE reactors could not be brought into service and were abandoned in 2008, after an enormous expenditure of funds. For the foreseeable future radioisotopes will continue to be produced by NRU.

Chalk River and Whiteshell were at the forefront of developing accelerators, which are widely used in research and have begun to replace gamma-ray machines for cancer therapy and sterilization of medical supplies. Accelerators are not nuclear devices; they are electrically driven and produce beams of high-energy particles such as electrons, which are used for a large variety of industrial processes. These include the production of heat-shrinkable plastics, the printing of beer-can labels, and the sterilization of medical supplies.

Chalk River and Whiteshell also became international centres of research on zirconium and its alloys, materials that are vital to CANDU reactors.

Chalk River tried to augment its shrinking government funding with non-nuclear commercial research work. The program has had a few successes, including a high profile O-ring design for the American space-shuttle program (see sidebar). Another application is the detection of metal particles in motor oil using non-intrusive eddy-current tests. This allows the oil in the engines of car and truck fleets to be readily tested so they can be serviced when necessary.

In 1962 AECL's Whiteshell Laboratories was established in eastern Manitoba; in its hey-day it had a staff of about 1,200. A large research program was carried out, much of it in collaboration with Chalk River. Whiteshell housed the WR-1 reactor, a novel organic-cooled, heavy-water-moderated reactor, which could operate at higher temperatures (350 to 400 °C/662 to 752 °F) and thus better efficiency, than the water-cooled CANDU reactor (300 °C/572 °F). However, the 60-megawatt (thermal) reactor, which used slightly enriched uranium

Trained Seals

In the early days of the CANDU program, pump seals in the primary coolant circuit of the reactors had to be replaced at least once a year because of the high pressure and temperature conditions in which they operated. To avoid these lengthy and costly reactor shutdowns, a research program was initiated to develop seals that would last at least five years. Using computer modeling as well as extensive laboratory testing, Chalk River staff gained a fundamental understanding of the mechanisms and material properties involved, and developed a seal that met the objectives. Power utilities across North America were experiencing similar problems, and in the late 1970s and early 1980s Chalk River won $18 million in contracts to solve pump-seal problems for three US nuclear electric utilities.

But the highlight of this research came after the tragic explosion of the US space shuttle Challenger in 1986, in which seven lives were lost. The cause was traced to hydrogen fuel leaking past an O-ring seal at the base of the booster rocket. The development of new O-rings was given top priority by NASA, and AECL was hired as a specialist subcontractor to advise on the re-design and to perform tests. In recognition of their efforts, two Chalk River scientists were invited to attend the launch of the first space shuttle with the new O-ring. No problems have been encountered since that time.

fuel and was commissioned in 1965, was shut down in 1985 and the research discontinued after the CANDU proved to be successful and budgets became tight.

Canada's research into the deep underground disposal of spent nuclear fuel from reactors was coordinated from Whiteshell. A major focus is at a nearby rocky granite outcrop, where the shaft for the Underground Research Laboratory penetrates into the Canadian Shield. A series of experiments have been undertaken to simulate the deep disposal of spent nuclear fuel (see Chapter 10).

Whiteshell is currently operating with a pared-down staff of about 250 who are decommissioning the site and the Underground Research Laboratory.

Nuclear Research in Universities

A number of Canadian universities conduct nuclear research using small research reactors, primarily the SLOWPOKE (see Figure 16-3).

Research reactors have a wide range of uses, including:

- neutron scattering in which beams of thermal neutrons are scattered by atoms in a target, yielding information regarding the atomic structure, magnetic state, and atomic binding energies of the target material;
- neutron activation analysis, a method for detecting very low concentrations of elements in a sample;
- neutron radiography (the neutron analogue of X-Rays);
- irradiation testing of materials;
- production of radioisotopes for research, medicine, and industry.

These reactors are used by researchers in an incredibly broad range of fields, such as archaeology, materials science, fusion research, mineral prospecting, and environmental science.

Research reactors are generally much smaller and simpler than power reactors with power levels usually only a fraction of one percent of a typical CANDU reactor. Their cores can often have very high power density, although contained in a very small volume.

The SLOWPOKE reactor has become an important research tool, beginning in 1971 when the first one was licensed at the University

Table 16-1 Canadian Research Reactors (non AECL)

University	Type	Thermal Energy
McMaster University	Swimming Pool	5,000 kW
University of Toronto	SLOWPOKE II	20 kW (closed 1998)
Royal Military College	SLOWPOKE II	20 kW
Saskatchewan Research Council	SLOWPOKE II	20 kW
University of Alberta	SLOWPOKE II	20 kW
Ecole Polytechnique	SLOWPOKE II	20 kW
Dalhousie University	SLOWPOKE II	20 kW

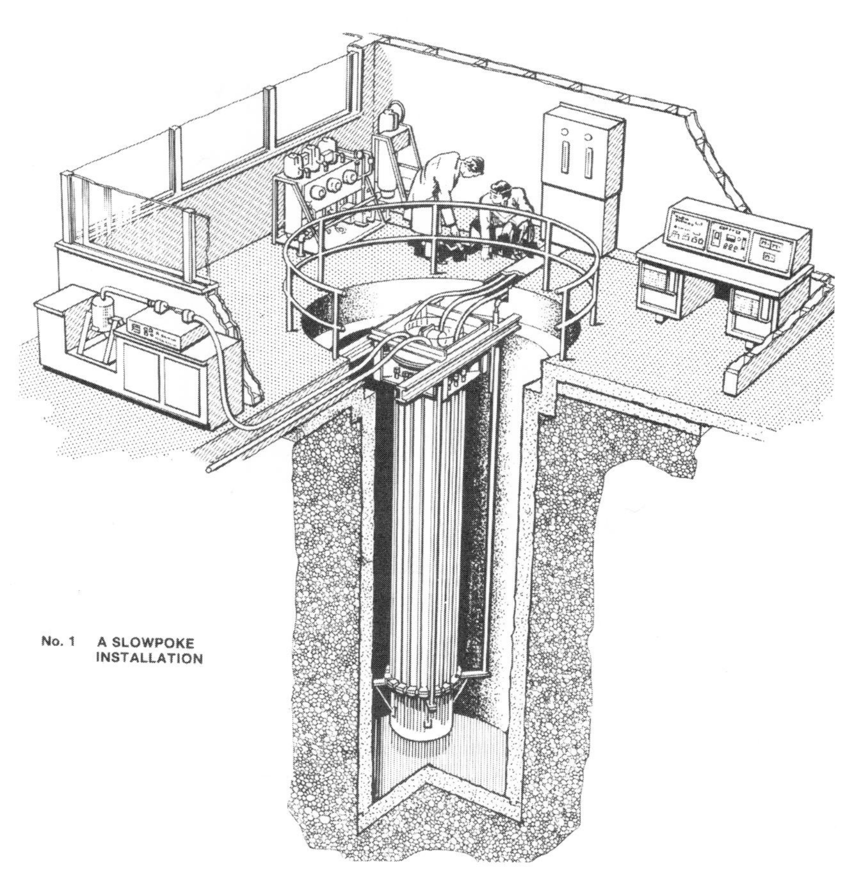

FIGURE 16-3: Schematic of a SLOWPOKE II reactor showing its simplicity. This highly versatile research reactor can operate unattended for long periods of time.

of Toronto. In 2008 four SLOWPOKE reactors were in operation at Canadian universities and one at the Saskatchewan Research Council. (Another is at the University of West Indies in Jamaica.) The SLOWPOKE at the University of Toronto was closed in 1998.

The key feature of the SLOWPOKE is that it is inherently safe and can run unattended. It is a small reactor with a fuel assembly about the size of a 4-litre can of paint. The core is located at the bottom of a cylindrical tank of ordinary water. The fuel initially consisted of highly enriched uranium (about 90 percent uranium-235) that generated only a small amount of heat, about 20 kilowatts. The highly enriched fuel in all these research reactors has been replaced by low-enriched uranium, about 20 percent uranium-235. The small size of the core is possible because it is surrounded by beryllium, which reflects neutrons back into the core. The heated water rises to the top of the tank where it loses its heat by flowing through a heat exchanger. The cool water sinks to the bottom of the tank, completing the natural convection cycle.

The SLOWPOKE is very simple in design as no pumps are needed and the water is not pressurized as in a CANDU reactor. Because the depth of water provides good shielding, the top can be kept open to allow easy access for monitoring, equipment, and experiments.

Due to the inherent passive safety, no elaborate, fast-acting shutdown systems are needed. As the temperature around the core rises, the water becomes less dense and is less efficient as a moderator, automatically causing the nuclear reaction to slow down. Any extra heat acts as a regulator and slows down the reaction. The SLOWPOKE has been designed so that this self-regulation can cope with all possible power variations, making it truly safe to operate.

The McMaster Nuclear Reactor, the oldest and largest university reactor in the British Commonwealth, was opened in 1959 at McMaster University in Hamilton, Ontario. Since then, it has provided facilities for neutron beam experiments, isotope production, neutron activation research, neutron radiography research, and education in many fields including materials science, nuclear science and engineering, and health and radiation physics. The reactor is a 5-megawatt (thermal) pool-type reactor of the Materials Testing Reactor design. Although relatively common in the United States, this is the only such reactor in Canada. It provides about 250 times more power than the SLOWPOKE reactor.

The McMaster reactor contributes to aircraft safety by performing radiographs of high-temperature jet engine turbine blades to detect structural flaws. It also produces about 25 percent of the world's supply of iodine-125, an isotope used in nuclear medicine for the treatment of prostate cancer. The fuel, which consists of uranium-aluminum plates enriched to about 20 percent uranium-235, is clad in aluminum and sits in a 10-metre-deep pool filled with water.

In 1999, the Canadian Foundation for Innovation, the Ontario Innovation Trust, industry, and the university together contributed $10

The Dalhousie Reactor

The SLOWPOKE reactor at Dalhousie University in Halifax, Nova Scotia, has thrust its Trace Analysis Research Centre into the forefront of research in analytical chemistry. Installed in 1976, the reactor is used for Neutron Activation Analysis. In this method a sample is irradiated in the reactor, where the nuclei absorb some neutrons, become unstable—in other words, radioactive—and emit gamma rays. These can be measured with a gamma-ray detector. This yields a gamma energy spectrum, which can be compared to the spectra of known elements to identify the elements in the sample.

It is imperative that the samples be removed from the reactor and placed in a detector before the elements with very short half-lives decay away. The scientists at Dalhousie have developed a system that does the transfer in 0.35 seconds. The reactor is used to study, among many other things, how trace elements are bound to proteins in food and in the body. The SLOWPOKE is also used to produce isotopes for medical research.

million to refurbish the McMaster reactor and support new research programs. This capital injection, together with the revenues from its commercial isotope production, has ensured its continued operation for many years.

The Need for Neutrons

In the late 1990s, governments in Canada recognized that research facilities in universities were run down and needed upgrading. They also recognized that research and the innovative technologies it produces are the key to the growth of Canada's increasingly knowledge-based economy. For these important reasons, billions of dollars are being invested in improving research infrastructure in Canadian universities. In parallel, up to 2,000 new research professorships are being funded. These programs are producing an upsurge of activity in research of all kinds.

Nuclear research and development in Canada, however, has been in decline since the halcyon days of the sixties and seventies. This decline became precipitous during the mid-1990s when deficit reduction measures greatly reduced federal funding at the same time as nuclear utilities decreased their research budgets. Recruiting new graduates for the nuclear industry has become increasingly difficult at a time when retirements are rapidly depleting the work force and a nuclear renaissance is gaining momentum.

The eventual closure of the NRU reactor will leave a major gap in Canada's nuclear research capability. The existing university reactors cannot provide the high neutron fluxes and research infrastructure available at a national laboratory. The National Research Council and AECL, in partnership with universities and industry, have been lobbying for years to replace the aging NRU reactor with a new research reactor called the Canadian Neutron Facility. It would provide advanced materials research capability for universities and industry, helping maintain Canada's competitiveness in this field. In addition, it would provide the focal point for a new generation of nuclear researchers.

Neutrons produced by reactors are unique, deep-penetrating, and non-destructive probes that can be used to study a variety of materials including pharmaceuticals, magnetic devices such as computer disks, polymers, welded structures, superconductors, biological materials, and much more. In addition, only neutrons can detect residual stress deep inside alloys and ceramics and so contribute to the reliability and safety of new industrial products.

About one hundred Canadian and international scientists and engineers, as well as about 20 graduate students, currently conduct research at NRU. This number is expected to more than triple if the Canadian Neutron Facility comes online.

To help put the Canadian Neutron Facility in context, it can be compared to another research facility, the Canadian Light Source, completed in 2004 near Saskatoon, Saskatchewan. This is a type of electron accelerator called a synchrotron that accelerates a fine beam of electrons to very near the speed of light in a race-track-shaped accelerator about the size of a football field. At such high speeds, very bright X-ray beams are produced. These beams will be used to probe a variety

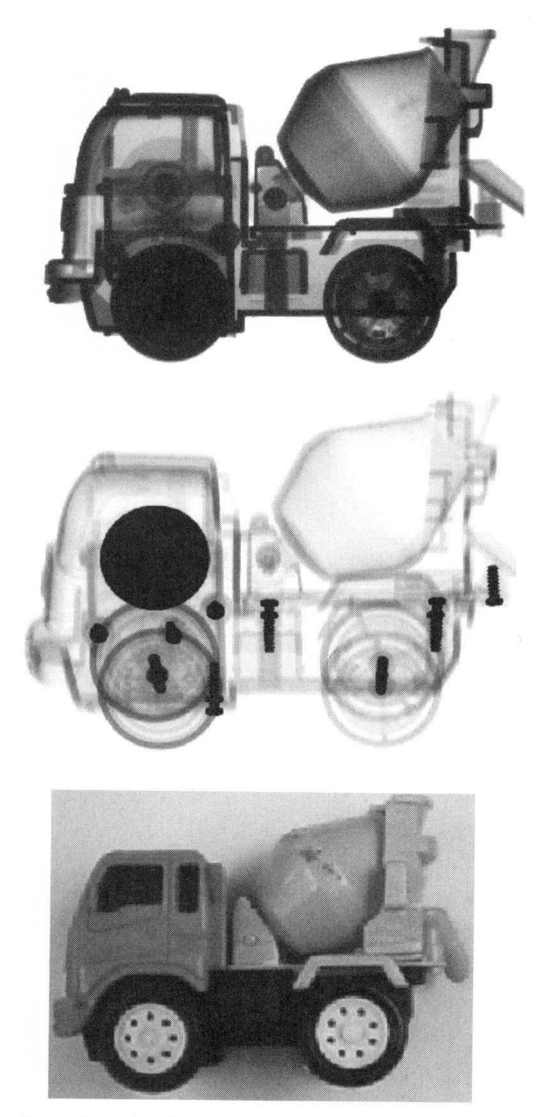

FIGURE 16-4: Imaging a toy truck using normal photography (bottom panel), neutrons (top panel), and X-rays (middle panel). Note that the X-rays show metal components very well while the neutrons reveal other structures not seen by X-Rays.

of materials including biological molecules and tissues. The Canadian Light Source is a necessity for modern materials research in Canada.

The simple case of the toy truck illustrated in Figure 16-4 shows why it is also essential to have a modern neutron source. The reason is that neutrons and synchrotron radiation "see" materials differently, thereby complementing each other. Researchers need both: it is most effective to investigate the various aspects of materials using both neutrons and X-rays.

To continue to advance innovation in many areas, Canada will need a new research reactor; if not, it will again find itself behind other countries. For example, an advanced high neutron flux materials test reactor, the Jules Horowitz Reactor, is under construction at Cadarache, France funded by a consortium of many countries and nuclear firms (including AREVA). The Canadian Neutron Facility would ensure the survival of a national laboratory where ingenuity will continue to transform research and development into innovative technological applications for the benefit of all Canadians.

GLOSSARY

Absorption: A type of nuclear reaction in which a nucleus captures a neutron.

Abundance: In the context of an isotope, the relative occurrence of that particular isotope compared to others of the same element.

Actinides: A group of 15 elements with atomic number from 89 (actinium) to 103 (lawrencium). All are radioactive.

Activation: The process whereby a non-radioactive (stable) atom becomes radioactive (unstable) due to a change in the nucleus, usually caused by exposure to external radiation.

AECB: Atomic Energy Control Board, the former name of the CNSC (see below).

AECL: Atomic Energy of Canada Limited, the federal Crown corporation responsible for nuclear research and development and CANDU reactor sales and service.

ALARA: As Low As Reasonably Achievable, a philosophy of nuclear safety in which risks should be maintained not just at regulatory limits, but as low as reasonably achievable with consideration to economic and social conditions.

Alpha particle (α): A positively charged particle emitted in the radioactive decay of certain radioactive atoms. An alpha particle is identical to the nucleus of a helium atom, consisting of two neutrons and two protons.

Atom: The basic building block of matter. It is the smallest unit of a material that retains all of the properties of that material. It consists of a nucleus surrounded by electrons.

Atomic number: The number of protons in the nucleus of an atom (equal to the number of orbital electrons). This determines which chemical element an atom is. For example, an atom with six protons is carbon.

Atomic weight: The number of protons and neutrons in the nucleus of an atom expressed in grams.

Becquerel (Bq): The unit for expressing the rate of radioactive decay. One becquerel is equivalent to one disintegration per second.

Beta particle (**β**): A charged particle emitted from the nucleus of certain radioactive (unstable) atoms. If negatively charged it is an electron, if positively charged it is a positron.

Breeder reactor: A reactor that produces more fissile material (fuel) than it consumes.

Burn-up: The amount of energy produced by a given amount of nuclear fuel.

BWR: Boiling Water Reactor, a type of light water reactor that generates steam directly in the reactor vessel.

Calandria: The vessel that forms the core of a CANDU reactor. It contains the heavy-water moderator and is traversed by horizontal tubes containing the fuel bundles and through which the coolant flows.

Cancer: A disease in which cells multiply uncontrollably in the body and interfere with the normal function of the organism.

CANDU: Canada Deuterium Uranium, a Canadian nuclear power reactor system using deuterium as moderator and natural uranium as fuel.

Chain reaction: A sequence of nuclear fission reactions in which the neutrons produced by the fissioning of one nucleus initiate fission in one or more other nuclei.

Cladding: The metal surrounding the uranium fuel in a fuel bundle. For CANDU fuel the cladding is a zirconium alloy.

CNF: Canadian Neutron Facility, a proposed new experimental reactor to replace NRU (see below).

CNSC: Canadian Nuclear Safety Commission, the federal body responsible for the regulation of nuclear activities in Canada.

Control rods: Rods consisting of a neutron absorbing material that are inserted into a reactor to stop the chain reaction.

Conversion: In uranium processing, the step in which uranium oxide (U_3O_8) is chemically converted to uranium hexafluoride (UF_6).

Coolant: A liquid or gas used to remove the heat from the core of a nuclear reactor. The most commonly used coolants are light water and heavy water.

Cosmogenic: Radioactive materials produced by cosmic radiation interacting with elements in the upper atmosphere.

Core: The central part of a nuclear reactor where the fuel is located and the chain reaction occurs.

Curie (Ci): The (old) unit for expressing the rate of radioactive decay. 1 curie = 3.7×10^{10} disintegrations per second. One curie is the radioactivity of one gram of radium.

Critical: In the nuclear context, this describes a condition where a chain reaction is sustained.

Critical mass: The amount of a fissionable material required to just sustain a chain reaction. Its value depends on a variety of factors, for example, the presence of a moderator and the geometry of the arrangement of the material.

Daughter: Decay product. An older term used to describe the new nucleus that results when a given nucleus decays.

Decommissioning: The process of permanently removing a reactor from service which involves the safe dismantling of storage of its radioactive components.

DNA: Deoxyribonucleic Acid, long-chained double-helix molecules that contain the genetic information of an organism. They are found mostly in the nucleus of cells.

Deuterium: The isotope of hydrogen, hydrogen-2, that has a nucleus containing one proton and one neutron. Also called heavy hydrogen, deuterium nuclei take the place of hydrogen nuclei in heavy water.

Dose: General term for the quantity of radiation received, also called exposure.

Electrolysis: The process of breaking up water molecules into hydrogen and oxygen by applying electricity.

Electromagnetic spectrum: The complete range of frequencies of electromagnetic waves from the lowest to the highest, including radio, infrared, visible light, ultraviolet, X-ray, gamma ray, and cosmic ray waves.

Electron: A subatomic particle with a negative charge that exists in the electron cloud surrounding the nucleus of an atom.

Element: A substance with atoms all of the same atomic number (same number of protons and electrons), but possibly different atomic weight (different isotopes).

Enriched uranium: Uranium in which the concentration of the isotope uranium-235 has been increased to greater than the natural concentration of 0.7 percent by weight.

Fast Reactor: A type of nuclear reactor without a moderator where the fission takes place at high neutron energies. Usually a fast reactor is a breeder reactor.

Fertile: An element that can be transformed by nuclear reactions into a fissile element.

Fissile: An element whose nucleus can be fissioned (split) by being stuck by a neutron.

Fission: The splitting of a nucleus into smaller nuclei, usually two nearly equal ones. This is accompanied by the emission of neutrons and a significant amount of energy. Fission in a reactor is initiated by bombarding fissionable material, usually uranium-235 with neutrons.

Fossil fuel: A fuel such as oil, coal, or natural gas originating from ancient vegetation which is burned to produce energy.

Fusion: A process in which two or more light nuclei combine to form a heavier nucleus with the release of energy. Fusion is the source of energy for the sun and the stars.

Gamma ray (γ): Energetic electromagnetic radiation (also called a photon) emitted by the nuclei of some radioactive elements due to nuclear transitions.

Genetic effect: DNA mutation that can be transmitted to the next generation (offspring).

Gray: A unit of radiation dose, 1 joule/kilogram.

Half-life: The amount of time needed for half of the atoms in a quantity of a radionuclide to decay.

Heavy water: Water in which the natural concentration of deuterium, about 0.015 percent or one in 6,600, has been increased to a high value, typically more than 95 percent. Heavy water is the moderator in CANDU reactors.

HLW: High-Level (Radioactive) Waste.

Hormesis: A theory that small exposures to radiation have positive effects on biological systems.

IAEA: International Atomic Energy Agency, a United Nations agency with the purpose of promoting the peaceful uses of nuclear technology and preventing its diversion to weapons uses. It manages the world nuclear safeguards system.

Ion: Atom, molecule, or molecular fragment carrying a positive or negative electrical charge.

Ionizing radiation: Radiation that has enough energy to remove electrons from neutral atoms or molecules that it passes through, creating ions.

Irradiator: A device for exposing materials to radiation for a variety of scientific and industrial purposes.

Isotopes: Atoms of the same element, that is, having equal numbers of protons, but with different numbers of neutrons. Isotopes of the same element have the same chemical properties, because they have the same number of electrons.

Light water: Ordinary water as found in nature, compared to heavy water.

LLRW: Low-Level Radioactive Waste.

LWR: Light-Water Reactor, the most common type of nuclear reactor, which uses light (ordinary) water as a moderator. There are two types of LWR: the boiling water reactor (BWR) and the pressurized water reactor (PWR).

MAPLE: Multipurpose Applied Physics Lattice Experiment, a new type of isotope production reactor.

Mass number: The total number of protons and neutrons in the nucleus of an atom.

Megawatt (MW): One million watts, the usual unit used to express the electrical output of an electrical power plant, such as a nuclear reactor. One watt in the SI system of units is one joule per second.

MOX: Mixed Oxide Fuel, reactor fuel consisting of a mixture of uranium and plutonium.

Moderator: The material used in a nuclear reactor to slow down, i.e., moderate, fast-moving neutrons so they will interact with the nuclei of fissile isotopes such as uranium-235 causing them to split, i.e., fission.

Molecule: The smallest unit into which a chemical compound containing one or more atoms can be divided and still maintain its chemical characteristics.

Natural uranium: Uranium with the isotopic abundance found naturally, i.e., 0.7 percent uranium-235 with the remainder uranium-238.

Neutron: A subatomic particle with no electrical charge and a mass nearly equal to that of a proton. It occurs in the nucleus of all atoms except hydrogen.

Neutrino: A highly energetic particle with a very low mass produced in many nuclear reactions in the universe.

Noble gases: The chemically inert gases helium, neon, argon, krypton, xenon, and radon.

NPD: Nuclear Power Demonstration, a 22-megawatt reactor that was the first in Canada to generate electricity (in 1962).

NRU: National Research Universal, a powerful research and isotope production reactor located at Chalk River.

NRX: National Research Experimental, the first large research reactor at Chalk River, operating from 1947 to 1992.

Nuclear fuel cycle: All the steps necessary to use uranium to produce electricity including mining and milling of uranium, converting the uranium to a fuel, fissioning the uranium in a reactor to create power, and disposing of the waste. This term can apply more generally to other reactor fuelling schemes involving thorium, plutonium, and other elements.

Nuclear radiation: Ionizing radiation (alpha, beta, gamma, neutrons) emanating from the nucleus of radioactive atoms as they undergo radioactive decay to a more stable form.

Nuclear reactor: A device in which a nuclear chain reaction is induced for power production, or other industrial and scientific purposes.

Nuclides: A general term used for species of atomic nuclei; radionuclides are radioactive nuclides.

OPG: Ontario Power Generation, Canada's largest nuclear utility.

PET: Positron Emission Tomography, a medical diagnostic technique that employs the gamma rays that are created by the positrons emitted from the decay of certain radioactive nuclei.

Photon: A discrete amount of electromagnetic energy, or a quantum of light.

Plate tectonics: The study and description of the motions of the large plates which comprise the earth's surface.

Plasma: A gas consisting of electrically charged atoms and electrons.

Poison: Name (jargon) given to a substance that absorbs neutrons well, for example boron and gadolinium.

Positron: A beta particle with a positive electrical charge.

Ppm: Parts per million.

Primordial radionuclides: The radioactive nuclei formed at the birth of the universe.

Proton: A subatomic particle in the nucleus of an atom with approximately the same mass as a neutron and carrying a positive charge equal but opposite to that of the electron.

PWR: Pressurized Water Reactor, the most common type of electricity producing reactor and one of the two general types of light-water reactors.

Radioactive decay: The spontaneous emission of radiation from the nucleus of a radioactive atom. It is the process by which a radioactive (unstable) nucleus becomes stable.

Radioisotope: An isotope (see above) that is radioactive; in other words, an isotope that undergoes radioactive decay.

Radiopharmaceutical: A drug containing one or more radioactive elements.

Reprocessing: The chemical processing of spent reactor fuel to recover the fissile and fertile material present in it.

Safeguards: A system of international treaties with monitoring designed to prevent nuclear materials from being diverted for weapons.

Sievert (Sv): A unit of exposure of a living organism to ionizing radiation. It is a measure of the health risk of ionizing radiation on human tissue.

SLOWPOKE: A low-power research reactor designed by AECL that is safe for unattended operation.

SNO: Sudbury Neutrino Observatory, an experimental apparatus installed in a mine in Sudbury, Ontario used to measure cosmic neutrinos, particularly those emitted by the sun.

Solar wind: The stream of particles emitted by the sun.

Somatic effect: Effect of radiation on the exposed individual and not on the offspring (genetic effect).

SPECT: Single Positron Emission Computerized Tomography, a more sophisticated version of PET.

Spent fuel: Nuclear fuel that has been used in a nuclear reactor to the point where it can no longer produce economic power. Also known as irradiated fuel or used fuel.

Tailings: The rock remaining after a mineral has been extracted.

Thermal reactor: A reactor in which the neutrons are moderated (slowed down) to permit fission at thermal (low) neutron energies.

Tokamak: A type of magnetic-confinement fusion reactor that uses current generated in a plasma to confine the plasma.

Transuranic: Elements having an atomic number greater than 92 (uranium). Transuranic elements are artificially produced and are radioactive.

TRIUMF: Tri University Meson Facility, a cyclotron (a type of particle accelerator) at University of British Columbia.

Tritium: Hydrogen-3, a naturally occurring radioactive isotope, also created in heavy water during reactor operation.

Wilson Cloud Chamber: A device for viewing particles by observing the ionization tracks they make in a vapour.

X-ray: Electromagnetic radiation used in medical and dental diagnoses because it penetrates matter. Although very similar to gamma rays, X-rays are created by energy transitions in the electron cloud that surrounds the nucleus.

ZEEP: Zero Energy Experimental Pile, the first nuclear reactor built outside the United States; it operated at Chalk River from 1945 to 1970.

Appendix A

The Basics of Radiation

The Atom and its Nucleus

To understand nuclear radiation, we must consider the basic building block of all matter: the atom. An atom is the smallest unit of a chemical element that retains all the properties of that element. The word atom comes from the Greek word *atomos*, meaning indivisible. At one time, scientists thought that the atom was the smallest unit of matter.

Our solar system occupies a huge volume in space, most of which is empty. The atom is similar. Far inside a cloud of negatively charged electrons is a tiny positively charged nucleus. For a different perspective, consider a baseball lying on the ground in the middle of a large stadium. If an atom were blown up in scale many billions of times, the dome would represent the outer fringe of electrons and the baseball would represent the atom's nucleus.

This tiny nucleus is what nuclear power plants use to provide us with electricity.

Now let us zoom in even closer and look at the nucleus.

Scientists have found that the nucleus is a bundle of two types of particles. The first type is called a proton and carries a positive charge. This charge is equal but opposite to the negative charge of an electron. An element is defined as an atom with a specific number of protons in its nucleus. There are 92 chemical elements that occur in nature and another dozen elements that have been created in laboratories. The number of protons is a unique characteristic of an atom, distinguishing which element it is. For example, an atom of oxygen always has 8 protons, gold always 79, lead 82, uranium 92, and so on. So the 92 natural elements are uniquely identified by having from 1 to 92 protons.

Neutrons, the second type of particle in the nucleus, are electrically neutral, as the name suggests. Neutrons and protons have almost the same mass, and this is more than 1,800 times greater than the mass of an electron. The nucleus of a specific element always has the same number of protons, but it can have different numbers of neutrons. Each element, then, has a "family" of atoms—called isotopes—whose nuclei contain different numbers of neutrons. For example, hydrogen has three isotopes: each has one proton, which may be accompanied by zero, one, or two neutrons (see Figure A-1).

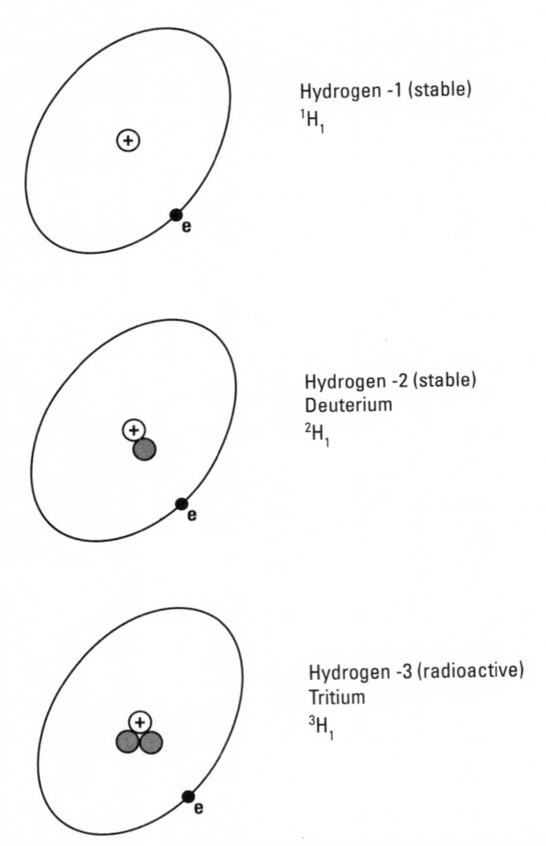

FIGURE A-1: Schematic of the three isotopes of hydrogen. Each isotope has one positively charged proton as indicated by the plus sign. Each also has one electron (designated by the e) orbiting the nucleus. Hydrogen-2, also called deuterium, has a neutron in its nucleus in addition to the proton. Hydrogen-3, also called tritium, has two neutrons. Tritium is radioactive and decays to helium-3 by emitting a beta particle.

Under normal circumstances, the number of electrons orbiting around the nucleus is the same as the number of protons in the nucleus. Because these are the same, the atom has no overall charge. The chemical behaviour of an element is determined by its electrons.

This is why iron is so different from oxygen, and gold from aluminum, and so on. The periodic table lists the elements in rows and columns: each column has similar chemical properties.

When a chemical reaction takes place between atoms—say, hydrogen and oxygen combining to form water—these reactions involve their electrons.

"Nuclide" is the general name for a nucleus containing a particular number of neutrons and protons. A chart of the nuclides can be constructed by plotting the known nuclides on a grid with the number of protons along the horizontal axis and the number of neutrons along the vertical axis (in a similar way as the periodic table arranges elements). In 2003, there were about 3,100 known nuclides.

To distinguish the different isotopes of an element, scientists have devised a system of notation. First, there is the the atomic mass number (often just called the mass number), which is the sum of the number of neutrons and protons. Second is the atomic number. This is the number of protons, which identifies the element itself. Uranium, for example, has 92 protons, and a naturally occurring isotope with 143 neutrons. This can be denoted as:

$$\text{Mass Number} \longrightarrow {}^{235}\mathrm{U}_{92} \longleftarrow \text{Atomic Number}$$

The superscript is the mass number: the number of protons plus neutrons (92 + 143 = 235), and this identifies the specific isotope. The subscript is the atomic number; this identifies the element itself.

Note that the atomic number (92) is redundant since the "U" means uranium which is defined as the element with 92 protons in its nucleus. Another common way of writing this is uranium-235 or U-235, since the name uranium identifies the element and thus the number of protons.

Neutrons, being electrically neutral, play no role in chemical reactions. For this reason, different isotopes of a given element always have the same chemical properties because they have the same number of electrons. However, they can have different nuclear properties. An element of particular interest is uranium, which in nature consists of a mixture of two isotopes; about 99.3 percent is uranium-238 (with 146 neutrons) and 0.7 percent is uranium-235 (with 143 neutrons).

Nuclide Stability and Radioactive Decay

Since a positive charge will repel a positive charge, and the same for two negative charges, why is it that positively charged protons in the nucleus do not fly apart? The reason is that the nucleus is held together by a very powerful localized force called the strong nuclear force. This is stronger than the electrical force, but it acts over a smaller distance. Neutrons help balance these opposing forces. If a nucleus has too many or too few neutrons to maintain this balance, it becomes unstable (or radioactive), and begins to follow a process called radioactive decay, also known as radioactive disintegration.

Quantum Mechanics

The world of the atom is governed by a physical theory known as quantum mechanics, developed in the early 1900s. It has been highly successful in explaining and predicting many atomic and nuclear phenomena, and it is now firmly established as a primary theory in physics. Understanding quantum effects, however, can be difficult given that they are often not what we would expect from our everyday experience. Some of the concepts from quantum mechanics run counter to our normal experience: subatomic objects exist as both waves and particles; the very act of observing these objects can change their properties (so there will always be a degree of uncertainty); and objects are distributed in space according to a system of probabilities. The good news is that you don't have to know about quantum mechanics other than to be aware we are making great simplifications in this Appendix for purposes of clarity.

Many nuclides are stable and will remain that way forever. Nuclear radiation, however, comes from those nuclei that are unstable—that is, radioactive. Most nuclei are unstable. What does this mean? One way of looking at this is to consider that a nucleus is stable for certain neutron numbers and unstable for others. A chart of the different nuclides reveals that, in general, the number of neutrons in stable nuclides is somewhat greater than the number of protons. This is especially true for the heavier elements. The pattern seen in the chart is that each element only has a limited number of isotopes (up to about 20 at most) with neutron numbers in a range around those of its stable isotopes. Of the approximately 3,100 known nuclides only 290 are stable, the remainder are radioactive. Radioactive nuclides, or radionuclides, have excess nuclear potential energy and move toward stability by giving off energy; in other words, they emit radiation.

The primary types of nuclear radiation that are released during the decay process were named by pioneering scientists after the first three letters of the Greek alphabet.

An "alpha" particle consists of two neutrons and two protons—essentially a helium nucleus. It is massive compared to other forms of radiation and, therefore, can penetrate only short distances into matter. For example, it is stopped by a piece of paper or by skin. An alpha emission transforms the emitting element to a new element; this new element has an atomic number that is two less. As well, the mass number decreases by 4. An example of alpha decay, representing the alpha particle by the Greek letter α, is:

$$^{238}U_{92} \longrightarrow {}^{234}Th_{90} + \alpha$$

In this nuclear reaction, a uranium-238 nucleus yields a thorium-234 nucleus and an alpha particle (α) when it decays. The remarkable thing is that one element transforms into another, a process called transmutation. The primary objective of early alchemists was to find a way of producing gold from other elements. We now know transmutation is possible only through nuclear reactions.

A beta particle (either an electron or positron) is emitted from inside the nucleus. An electron is formed when a neutron (n) converts to a proton (p) by emitting a nuclear electron (β) and gamma ray(s), denoted by γ:

$$n \longrightarrow p + \beta + \gamma$$

In this process the number of neutrons in the nucleus decreases by one and the number of protons increases by one. Again we have transmutation. An example of beta decay is:

$$^{60}Co_{27} \longrightarrow {}^{60}Ni_{28} + \beta + \gamma$$

The atomic number increases by one, but the total number of protons and neutrons stays the same. In other words, there is no change in the mass number. A beta particle can penetrate up to several centimetres into living tissue.

Positrons, denoted β^+, are beta particles that carry a positive charge. They are emitted from nuclei during the radioactive decay process when protons transform to neutrons.

$$p \longrightarrow n + \beta^+ + \gamma$$

An example of positron emission is sodium transmuting to neon:

$$^{22}Na_{11} \longrightarrow {}^{22}Ne_{10} + \beta^+ + \gamma$$

A positron is an "antiparticle"; it is seldom encountered because within microseconds it combines with an electron and both disappear, emitting two gamma rays. This phenomenon of positron annihilation is the basis of a diagnostic technique, positron emission tomography (see Chapter 11), that is widely used in nuclear medicine.

Gamma rays (γ), are electromagnetic radiation—similar to visible light, radio waves, microwaves, and X-rays—but emitted from nuclei. Unlike alpha and beta particles, gamma rays have no mass and no charge, only energy. Like all electromagnetic radiation, gamma rays travel at the speed of light. If only one or more gamma rays (and no other radiation) are emitted by the nucleus, then the change is called an "isomeric transition." This means that the atomic and mass numbers remain the same. The original, higher-energy-level nucleus is said to be in an "excited state" and is indicated by an "m" beside the mass number. An example of an isomeric transition is in the element technetium (a widely used radioisotope in medicine):

$$^{99m}Tc \longrightarrow {}^{99}Tc + \gamma$$

Gamma rays are similar to X-rays, but with higher energy. The primary difference is that gamma rays come from the nucleus, whereas X-rays arise from the atom's outer electron cloud. Like gamma rays, X-rays can penetrate relatively large distances into matter. Depending on their energy, gamma rays can penetrate up to a metre into concrete, and pass right through a human. It is important to stress that although X-rays cause the same health effects as nuclear radiation, they are an atomic effect, not a nuclear phenomenon.

Other transformations are possible inside the nucleus. An example of this is electron capture and internal conversion. Other particles may be emitted from the nucleus, such as neutrons, protons, and neutrinos. They will not be discussed further.

The stability of a nucleus depends largely on the mix of protons and neutrons. Except for hydrogen-1 and helium-3, the number of neutrons in a stable nucleus either equals or exceeds the number of protons. If an atom has too many neutrons, it decays by emitting a beta particle. In this process, a neutron is converted to a proton. If the nuclide has too many protons, it decays by emitting a positron or (rarely) by "capturing" an electron in the nucleus; here a proton is converted to a neutron. Decay by emitting an alpha particle occurs only in the heavy elements (atomic number 83—the element bismuth, and higher). In all these cases, a new daughter nuclide, or decay product, is created. In the example of alpha decay used above, thorium-234 can be said to be the daughter of uranium-238. If the daughter is unstable—i.e. radioactive—it will also undergo radioactive decay. This process will continue until a stable configuration is reached.

Radioactive Decay over Time

Each radioactive nucleus decays independently. That is, the nucleus emits alpha, beta, and/or gamma(s) at random times. Although any one atom behaves in a random manner in terms of when it decays, the average decay behaviour of a collection of many radioactive atoms is very predictable, and has been observed to always occur in an exponential manner.

A useful parameter for characterizing the rate of decay of a particular nuclide is its half-life. This is defined as the time it takes for half the atoms to decay to daughter atoms. In other words, this is the time for the radioactivity of the parent isotope to decrease by half. For example, the half-life of iodine-131 is 8.1 days. This means that if you started with 100 grams of iodine-131, after 8.1 days only 50 grams would be left (the other half would have transformed to another nuclide). After another 8.1 days, only 25 grams would be left, after another 8.1 days, only 12.5 grams would be left, and so on. If this is plotted on a graph it would take the form of an exponential curve, as shown in Figure A-2.

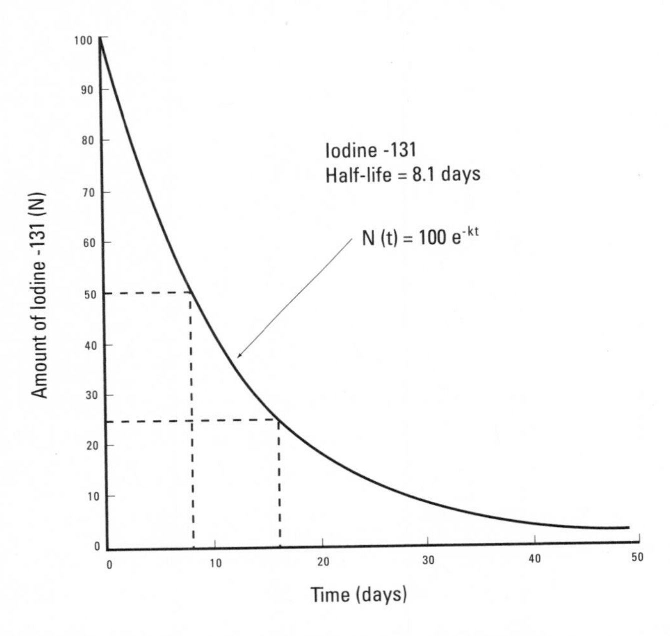

FIGURE A-2: Graph illustrating the concept of half-life for iodine-131. At the beginning there are 100 grams of iodine-131. After one half-life (8.1) half of them have decayed and only 50 grams remain. After 2 half-lives (16.2 days) only a quarter (25 grams) remain, and so on.

Half-life is an important concept. Every radioactive isotope or radionuclide decays following this exponential pattern until it becomes stable, that is, non-radioactive. Each radionuclide has its own distinctive half-life.

Some isotopes have very long-half lives. Uranium-238, for example, has a half life of 4.5 billion years. It is convenient to use scientific notation to write very large numbers or very small numbers and a brief explanation of this method is included at the end of this appendix.

Table A-1 lists the half-lives of some of the radionuclides that are discussed in this book. The first four nuclides listed in the table are naturally occurring isotopes found in virtually all the rocks that make up the earth. Because they have very long half-lives, these nuclides have been around since the earth was first formed (4.5 x 10^9 years ago). We can conclude that the earth was much more radioactive in the past.

Table A-1 Half-lives of some radionuclides

Nuclide	Half-life
Uranium-238	4.5 x 10^9 years
Uranium-235	0.71 x 10^9 years
Thorium-232	14 x 10^9 years
Potassium-40	1.3 x 10^9 years
Hydrogen-3 (tritium)	12 years
Carbon-14	5,730 years
Cobalt-60	5.3 years
Strontium-90	28 years
Cesium-137	30 years
Technetium-99m	6 hours

Radioactive Decay Chains

Uranium and thorium, two heavy radioactive atoms that occur naturally in rocks and soils, have the characteristic that they do not decay directly to a stable element. Instead, their daughter products are also radioactive; so in turn, the daughter decays into another daughter. This sequence continues through a chain of radioactive daughters until a stable one is reached. The uranium-238 decay chain is shown in Table A-2. Each isotope decays to the one below it in the table. For example, uranium-238 decays to thorium-234 (by emitting an alpha particle), and the half-life of this decay is 4.5 billion years. Then the thorium-234 decays to protactinium-234m ("m" means an excited state) by emitting a beta with half-life of 24 days, and so on. All the decays in the chain are happening simultaneously. It is seen that uranium-238 decays through 16 different daughters before reaching lead-206, which is stable.

Similar natural decay chains exist for uranium-235 and thorium-232. A neptunium series once existed but is no longer observed, as the longest half-life in the series is only 2.2 million years (this is small

Table A-2 The Uranium-238 Decay Chain

Isotope	Half-life
Uranium-238	4.5×10^9 years
Thorium-234	24 days
Protactinium-234m	1.3 minutes
Uranium-234	2.5×10^5 years
Thorium-230	7.7×10^4 years
Radium-226	1622 years
Radon-222	3.8 days
Polonium-218	3 minutes
Lead-214	27 minutes
Astatine-218	2 seconds
Bismuth-214	20 minutes
Polonium-214	1.6×10^{-4} seconds
Thallium-210	1.3 minutes
Lead-210	22 years
Bismuth-210	5 days
Polonium-210	138 days
Thallium-206	4.2 minutes
Lead-206	stable

compared to the age of our planet). In contrast, uranium-238 has gone through only one half-life since the earth was formed.

The complexity of these decay chains indicates how difficult it was for the early scientists to unravel them. It is a tribute to their patience and perseverance that they were able to accomplish this feat using what would now be considered primitive instruments.

Units of Radiation

To understand radiation in greater detail, we need to be able to talk about it quantitatively, that is, using numbers instead of generalities. To do this we must introduce the units in which radioactivity is measured.

The activity or amount of radioactivity in a substance is measured by the number of disintegrations—radioactive decays—that occur per second. One becquerel (Bq) is defined as one radioactive decay per second (this unit is named after Henri Becquerel, who in 1896 was the first to detect radioactivity). The concentration of radioactivity in gases, liquids, and solids is usually expressed in becquerels per cubic metre, litre, and kilogram, respectively. Table A-3 gives an idea of the amount of radioactivity found in natural substances.

Table A-3 Typical Amounts of Radioactivity in Natural Substances

Radon-222 in air	30 Bq/m^3
Radium-226 in drinking water	0.0003 Bq/L
Potassium-40 in soil	400 Bq/kg
Typical Human (total)	10,000 Bq

This table shows that there are typically about 30 radioactive decays per second caused by radon-222 in one cubic metre of air. Our bodies, for example, contain naturally occurring radioisotopes with an activity greater than 10,000 Bq.

The concept of radiation dose (defined as the radiation exposure received by a living organism), and its effect on living cells is discussed in more detail in Chapter 4. However, the unit used to measure dose is briefly introduced here.

When ionizing radiation penetrates matter, it transfers energy to the substance. The unit gray (Gy) is used to measure the absorbed energy per unit mass. One gray corresponds to the deposition of one joule of energy in one kilogram of matter. The gray is a large unit; more commonly the milligray (mGy) is used where the prefix "milli" means one thousandth—so one Gy contains a thousand mGy.

The unit used to measure the effect of radiation on humans is the sievert (Sv), named after the pioneering Swedish clinical physicist Rolf Sievert. The sievert is a measure of the energy per unit of mass deposited by radiation in the human body (expressed in Gy) times the likelihood of that causing damage. Thus the sievert takes into account the fact that different types of radiation cause different damage. For example, alpha particles cause more damage per unit of energy deposited than gamma rays. Doses received in real life are usually much smaller than a Sv, so the unit millisievert (mSv) is generally used.

The units described above are part of the internationally accepted system (Système International, SI) of units. The previous system of units

Table A-4 Typical Radiation Doses in millisieverts

Nuclear Power Plant Worker per year (above natural radiation)	0.1–1 mSv
Average Natural Radiation per year (Canada)	2.0 mSv
Typical Chest X-ray	0.3 mSv

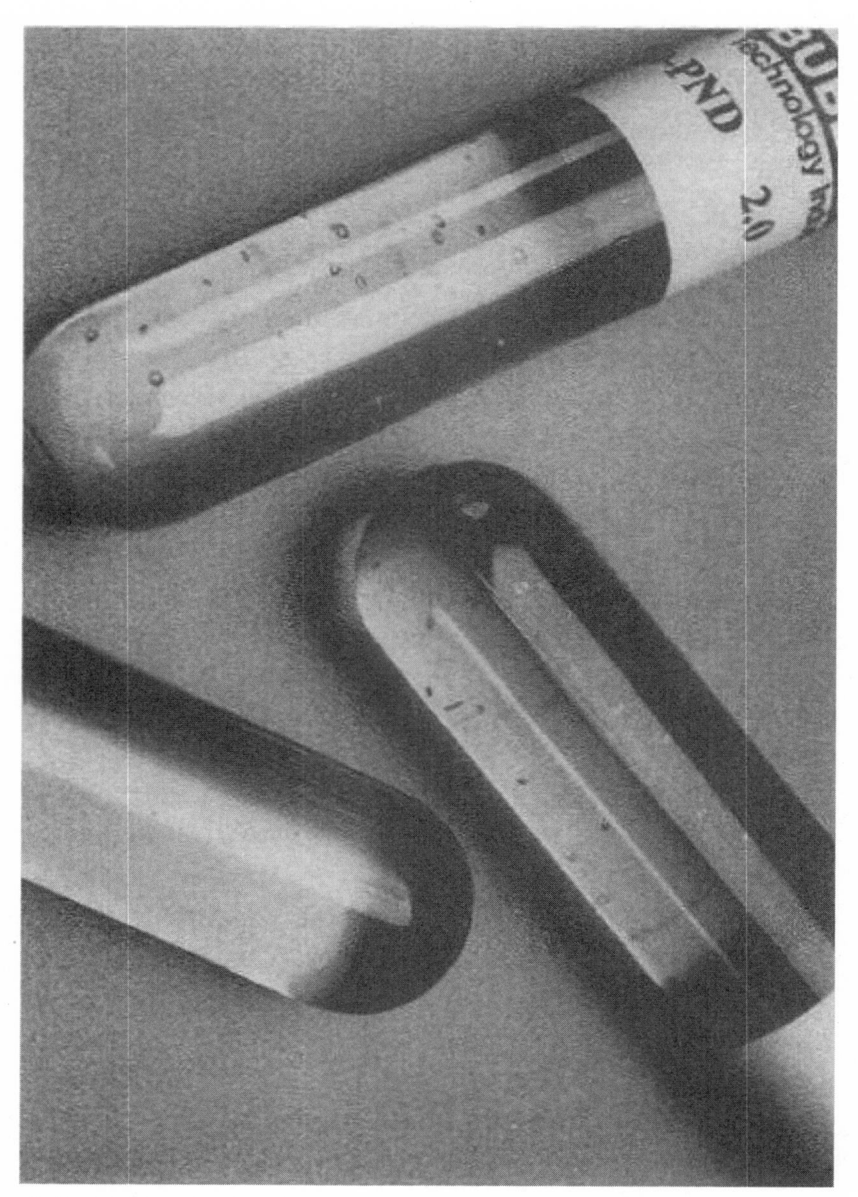

FIGURE A-3: Measuring the radiation from neutrons is difficult, but a Canadian company, Bubble Technology Industries, has come up with an ingenious solution. When neutrons strike a proprietary gel (lower tube without neutron exposure) they cause visible bubbles (upper two tubes). The bubbles are "frozen" in the gel and can easily be counted to determine the neutron field to which they were exposed. Bubble detectors have flown on many space missions and have been used in numerous other applications.

still occurs in the literature, although they will not be used in this book. The previous units and their equivalence to the SI system are as follows:

Activity:	1 curie (Ci) = 37 x 10^9 becquerel
Absorbed energy:	1 rad = 0.01 gray
Biological equivalent dose:	1 rem = 0.01 sievert

Detecting Radioactivity

Radioactivity is easy to measure because it can be detected at a distance; the main reason for this is the emission of gamma radiation. In comparison, the detection of non-radioactive substances requires sampling and chemical analyses at a laboratory, a much more costly and time-consuming process.

Some of the methods used to detect radiation are based on the fact that alpha and beta particles and gamma rays are all cause ionization. That is, they knock atomic electrons out of their orbits when they pass through matter. The atoms that have been struck are no longer neutral; they each have an electrical charge and are called ions. The best known radiation detector is the Geiger counter, a hand-held instrument that uses a high voltage to detect the small electrical currents caused by ionization, even down to a single ionizing particle. Geiger counters are used, for example, by prospectors to search for uranium deposits and by hospital safety workers to determine the extent of a radioactive spill (and of course to verify that the spill has been cleaned up).

A uniquely Canadian innovation for measuring neutron exposure, called the "bubble detector", is shown in Figure A-3. This ingenious but simple technology, discovered by Atomic Energy of Canada Limited and commercially developed by Bubble Technology Industries of Chalk River, Ontario, has gained world-wide acceptance.

Because of their inherent "detectability," radioactive materials are easy to study. In fact, the role of radioactive materials in the environment is generally far better understood than that of non-radioactive ones.

Scientific Notation

In science and engineering one often has to deal with very large and very small numbers. But too many zeroes can be confusing. For example, the half life of uranium-238 is 4.5 billion years. This can also be written as 4,500,000,000 years. The scientific notation for this number is 4.5 x 10^9 which means exactly the same thing but in shorthand. Similarly, we can write one thousand (1,000) as 10^3 or we can write one thousandth (0.001) as 10^{-3}. This becomes more convenient as the numbers involved become very large or very small.

This notation also makes it easy to multiply numbers and provides a neat method for keeping track of the "zeroes." For example, one hundred (10^2) times ten thousand (10^4) is one million (10^6). This result is easily obtained by adding the exponents on the 10s, in this case 2+4, which equals 6. This is particularly useful when there are many very large and very small numbers to be multiplied. For example:

0.022 x 11,000 = 2.2 x 10-2 x 1.1 x 104 = 2.42 x 102 = 242

Appendix B

Nuclear Web Sites

American Nuclear Society (USA): www.ans.org
Atomic Energy of Canada Limited: www.aecl.ca
Cameco Corporation: www.cameco.com
Canadian Nuclear Association: www.cna.ca
Canadian Nuclear FAQ: www.nuclearfaq.ca
Canadian Nuclear Safety Commission: www.nuclearsafety.gc.ca
Canadian Nuclear Society: www.cns-snc.ca
Hydro Quebec: www.hydro.qc.ca
International Atomic Energy Agency: www.iaea.or.at
International Radiation Protection Association: www.irpa.net
MDS Nordion: www.mds.nordion.com
Natural Resources Canada: www.nuclear.nrcan.gc.ca
New Brunswick Power Commission: www.gov.nb.ca/news/nbp
Nuclear Energy Institute (USA): www.nei.org
Nuclear Waste Management Organization: www.nwmo.ca
Radiation Safety Institute of Canada: www.radiationsafety.ca
Ontario Power Generation: www.opg.com
Society of Nuclear Medicine: www.snm.org
Uranium Information Centre (Australia): www.uic.com.au
Uranium Institute (UK): www.uilondon.org
Virtual Nuclear Tourist: www.nucleartourist.com
Wastelink: www.radwaste.org
  more than 6,500 links worldwide
World Nuclear Association (UK, formerly Uranium Institute):
  www.world-nuclear.org

Notes and Bibliography

Notes

5-1: "Severe Accident Analysis for Large Energy Systems," Paul Scherrer Institut, summarized by Colin Hunt, Canadian Nuclear Association, 2003. Online at www.cna.ca/english/pdf/Articles/ScherrerInstitut.pdf.

9-1: "Study Details Impact of Pollution on Public Health from Nine Older Fossil Fuel Power Plants in Illinois," Harvard School of Public Health, 2001. Online at www.hsph.harvard.edu/press/releases/press01032001.html.

11-1: Much of the statistics in Chapter 11 are taken generally from the web site of the Canadian Institute for Health Information at www.cihi.ca and specifically from Glenda Yeats, "Reading Between the Lines: A Closer Look at Today's Health Changes," Canadian Institute for Health Information, 2007.

11-2: Thirty million people undergo Tc-99m diagnostic procedures annually. Cristina Hansell, "Nuclear Medicine's Double Hazard: Imperiled Treatment and the Risk of Terror," *Nonproliferation Review*, 15.2: 2008. Online at cns.miis.edu/pubs/npr/vol15/152_hansell_nuclear_medicine.pdf.

11-3: Bone cancer and strontium-89. "Strontium-89 and Chemotherapy Curb Bone Cancer Pain," 2005. Online at www.urotoday.com/61/browse_categories/prostate_cancer/strontium89_and_chemotherapy_curb_bone_cancer_pain.html.

11-4: A summary of the NRU incident is provided in a statement by Minister of Natural Resources, Mr. Lunn. "Opening Statement by the Honourable Gary Lunn, P.C., M.P., Minister of Natural Resources," Natural Resources Canada, 2008. Online at www.nrcan.gc.ca/media/spedis/2008/200805-eng.php. The CNSC also provided a statement. "CNSC Statement Regarding AECL's NRU Reactor," Canadian Nuclear Safety Commission, 2007. Online at www.snm.org/index.cfm?PageID=7052&RPID=10.

11-5: Forty-thousand treatments per day. "The Nuclear Advantage," Keewatin Publications, undated. Online at www.cns-snc.ca/Publications/Nuclear_Advantage_Canada_Edition.pdf.

12-1: "Nuclear Facts: What are the health benefits of nuclear medicine?" Canadian Nuclear Association, undated. Online at www.cna.ca/english/pdf/NuclearFacts/Nuclear_Facts_Health_Benefits.pdf.

12-2: G. R. Malkoske et al., "Cobalt-60 Production in CANDU Reactors," undated. Online at www.nuclearfaq.ca/malkoskie_cobalt_paper.pdf.

Bibliography

Atomic Energy Control Board. 1995. *Canada: Living with Radiation.* Ottawa: Canada Communication Group.

——. 1998. *Radioactive Emission Data from Canadian Nuclear Generating Stations 1988 to 1997.* INFO-0210/Rev.8. Ottawa.

Atomic Energy of Canada Limited. 1994. *Environmental Impact Statement on the Concept for Disposal of Canada's Nuclear Fuel Waste.* AECL-10711, COG-93-1.

——. 1997. *Canada Enters the Nuclear Age: A Technical History of AECL.* Montreal and Kingston: McGill-Queens University Press.

——. 2007. *ACR-1000: Technical Summary: An Evolution of CANDU, AECL, Mississauga.* (Brochure.)

Ball, N.R. 1987. *Professional Engineering in Canada 1887 to 1987.* Ottawa: National Museum of Science and Technology/National Museums of Canada.

Blix, H. 1997. "Nuclear Energy in the 21st Century." *Nuclear News* 50.10: 34–39.

Bolton, R.A. 1999. "Fusion Research's Demise in Canada: Another Avro Arrow?" *Canadian Nuclear Society Bulletin* 20.3: 43–46.

Bothwell, R. 1984. *Eldorado: Canada's National Uranium Company.* Toronto: University of Toronto Press.

——. 1988. *Nucleus: The History of Atomic Energy of Canada Limited.* Toronto: University of Toronto Press.

Bushby, S.J., J.R. Dimmick, and R.B. Duffey. 1999. "Conceptual Designs for very High Temperature CANDU Reactors." *Canadian Nuclear Society Bulletin* 21.2: 25–31.

Calaprice, A. 1996. *The Quotable Einstein*. Princeton, NJ: Princeton University Press.

Canadian Environmental Assessment Agency. 1998. *Nuclear Fuel Waste Management and Disposal Concept*. Report of the Environmental Assessment Panel on the Nuclear Fuel Waste Management and Disposal Concept. Ottawa.

Carbon, M.W. 1997. *Nuclear Power: Villain or Victim? Our Most Misunderstood Source of Electricity*. Madison, WI: Pebble Beach Publishers.

Chen, W.L. et al. 2004. "Is Chronic Radiation an Effective Prophylaxis against Cancer?" *Journal of the American Physicians and Surgeons* 9.1: 6-10.

Canadian Nuclear Association. 2008. "Nuclear Canada Yearbook: 2008." Ottawa.

Cohen, B.L. 1990. *The Nuclear Energy Option: An Alternative for the 90s*. New York, NY: Plenum Press.

——. 1995. "Test of the Linear No-Threshold Theory of Radiation Carcinogenesis for Inhaled Radon Decay Products." *Health Physics* 68.2: 157–174.

Colditz, G.A., et al., eds. 1996. "Harvard Report on Cancer Prevention." *Cancer Causes and Control* 7 (Suppl.) S1-S55.

Cole, H.A. 1988. *Understanding Nuclear Power*. Aldershot, UK: Gower Technical Press.

Coursey, B.M., and R. Nath. 2000. "Radionuclide Therapy." *Physics Today* 53.4: 25–30.

Cousins, T. et al. 2000. "Using Thermal Neutron Activation to Detect Non-Metallic Land Mines." *Canadian Nuclear Society Bulletin* 21.2: 40–44.

Davies, J.A., G. Amsel, and J.W. Mayer. 1992. "Reflections and Reminiscences from the Early History of RBS, NRA and Channeling." *Nuclear Instruments and Methods in Physics Research*. Section B. 64: 12–18.

Duport, P. 2000. "The Effects of Low Doses of Ionizing Radiation." *Canadian Nuclear Society Bulletin* 21.3: 20–23.

Eisenbud, M. 1968. "Sources of Radioactivity in the Environment: Proceedings of a Conference on the Pediatric Significance of Peacetime Fallout." *Pediatrics* (Suppl.) 41: 174–195.

——. 1973. *Environmental Radioactivity.* 2nd ed. New York, NY: McGraw Hill.

Eldorado Resources Limited. 1985. *Uranium and Electricity*. 5th ed. Ottawa.

Ernst and Young Management Consultants. 1993. *Study of the Economic Benefits of the Canadian Nuclear Industry.* Conducted for Atomic Energy of Canada Limited. Toronto.

Foster, J. 2000. "In Memoriam: Harold Smith." *Canadian Nuclear Society Bulletin* 20.4: 43–45.

Freese, B. 2003. "Coal: A Human History." New York, NY: Penguin.

Fuller, J.G. 1975. *The Day We almost Lost Detroit*. New York, NY: Reader's Digest.

Gray, C. 1984. "Nuclear Energy and Medicine: The Canadian Connection." *Canadian Medical Association Journal* 130: 299–303.

Harms, A.A. et al. 2000. *Principles of Fusion Energy*. Singapore: World Scientific.

Hart, R.S. 1997. *CANDU Technical Summary*. Atomic Energy of Canada Limited.

Hedges, K.R., and S.K.W. Yu. 1998. "Next Generation CANDU Plants." *Canadian Nuclear Society Bulletin* 19.3: 16–20.

Hincks, E.P., ed. 1979. *Nuclear Issues in the Canadian Energy Context*. Ottawa: Royal Society of Canada and the Science Council of Canada.

Hore-Lacy, I. 1997. *Nuclear Electricity*. 4th ed. Melbourne, Australia: Uranium Information Centre Ltd.

House of Commons, Canada. 1988. *Nuclear Energy: Unmasking the Mystery.* 10th Report. Ottawa: Standing Committee on Energy, Mines, and Resources.

Hoyle, F., and G. Hoyle. 1980. *Common Sense in Nuclear Energy*. San Francisco, CA: W.H. Freeman.

Hurst, D.G. 1989. "The Road to CANDU." *Canadian Nuclear Society Bulletin*. 10.4: 17–21.

International Commission on Radiological Protection. 1990. "Recommendations of the International Commission on Radiological Protection." ICRP Publication 60. Pergamon Press.

Ing, H., et al. 1996. "Bubble Detectors and the Assessment of Biological

Risk from Space Radiations." *Radiation Protection Dosimetry* 65: 421–424.

International Atomic Energy Agency. 1997. *Ten Years after Chernobyl: What Do We Really Know?* Summary of the Proceedings of the IAEA/WHO/EC International Conference in Vienna. April 1996.

——. 1999. *The International Nuclear Event Scale*. Vienna.

——. 2000. "Nuclear Power Status around the World." *IAEA Bulletin* 42.3: 70.

Jackson, D.P. 1981. "Three Mile Island: a Personal Commentary." Hamilton: McMaster Institute for Energy Studies, *Energy Newsletter* 1: 4–13.

——and J. de la Mothe. 2001. "Nuclear Regulation in Transition: The Atomic Energy Control Board at the Turn of the Century." *Proceedings of the Conference on the Future of Nuclear Energy in Canada.* Toronto: University of Toronto Press.

Japan Atomic Industrial Forum. 1996. *Natural Radiations through Naked Eyes*. Nagoya, Japan.

Jones, N. 2001. "The Monster in the Lake." *New Scientist* March: 36–40.

King, A., and B. Schneider. 1991. *The First Global Revolution: A Report by the Council of the Club of Rome*. New York, NY: Pantheon.

Lambert, B. 1990. *How Safe is Safe? Radiation Controversies Explained*. London, UK: Unwin.

Lau, J.H.K. et al. 1999. "The Canadian CANDU Fuel Development Program and Recent Fuel Operating Experience." *Canadian Nuclear Society Bulletin* 20.3: 8–15.

Lidstone, R.F. 1996. "The Development of MAPLE Technology." *Canadian Nuclear Society Bulletin* 17.4: 32–38.

Lightfoot, H.D. et al. 2006. *Nuclear Fusion Fuel Is Inexhaustible*. Engineering Institute of Canada Climate Change Conference. IEEE. Ottawa.

Litt, P. 2000. *Isotopes and Innovation: MDS Nordion's First Fifty Years, 1946–1996*. Montreal and Kingston: McGill-Queens University Press.

Lochbaum, D. 1998. *The Good, the Bad and the Ugly: A Report on Safety in America's Nuclear Power Industry*. Cambridge, MA: Union of Concerned Scientists.

Marmorek, J. 1978. *Everything You Wanted to Know about Nuclear Power (But Were Afraid to Find Out!)*. Toronto: Energy Probe and the Pollution Probe Foundation.

McHughen, A. 2000. *Pandora's Picnic Basket: The Potential and Hazards of Genetically Modified Foods*. Oxford, UK: Oxford University Press.

McNeill, J.R. 2000. *Something New under the Sun: An Environmental History of the Twentieth Century World*. New York, NY: W.W. Norton.

Miller, G.T., Jr. 1997. *Environmental Science: Working with the Earth*. Belmont, CA: Wadsworth.

Mitchell, R.E.J. 1999. "Low Dose Effects: Testing the Assumptions." *Canadian Nuclear Society Bulletin* 20.2: 39–40.

Moore, B., and S. Guindon. 1999. Competitiveness of Nuclear Energy. *CRUISE: Conference on the Future of Nuclear Energy*. Toronto: University of Toronto Press.

Morrison, R.W. 1998. *Nuclear Energy Policy in Canada: 1942–1997*. Carleton Research Unit on Innovation Science and Environment. Ottawa: Carleton University.

Mosey, D. 2006. *Reactor Accidents*. 2nd ed. London: Nuclear Engineering International Special Publications.

Murray, R.L. 2001. *Nuclear Energy*. Boston, MA: Butterworth Heinemann.

National Research Council. 1990. *Health Effects of Exposure to Low Levels of Ionizing Radiation: BEIR*. V. Committee on the Biological Effects of Ionizing Radiation. Washington, DC: National Academy Press.

——. 1999. *Condensed-Matter and Materials Physics: Basic Research for Tomorrow's Future*. Washington, DC: National Academy Press.

——. 1997. *Canada's Energy Outlook: 1996–2020*. Ottawa.

——. 2000. *Energy in Canada 2000*. Ottawa.

Nuclear Energy Agency. 1995. *The Environmental and Ethical Basis of Geological Disposal of Long-lived Radioactive Wastes*. Paris.

——. 1998. *Nuclear Power and Climate Change*. Paris.

——. 2000. *Nuclear Education and Training: Cause for Concern?* Paris.

Ontario Hydro. 1988. *A Journalist's Guide to Nuclear Power*. Toronto.

Organizations United for Responsible Low-Level Radioactive Waste Solutions. 1994. *The Untold Story: Economic and Employment Benefits of the Use of Radioactive Materials. Washington, D.C.*

Paterniti, M. 2000. *Driving Mr. Albert: A Trip across America with Einstein's Brain*. New York, NY: Dial Press.

Pearce, F. 2000. "Kicking the [Petrol] Habit." *New Scientist* November: 34–40.

Perrow, C. 1984. *Normal Accidents: Living with High Risk Technologies*. New York, NY: Basic Books.

Poch, D.I. 1985. *Radiation Alert*. Toronto: Energy Probe/Doubleday Canada.

Pool, R. 1997. *Beyond Engineering: How Society Shapes Technology*. Oxford, UK: Oxford University Press.

Reeves, R. 2007. *A Force of Nature: The Frontier Genius of Ernest Rutherford*. New York, NY: W.W. Norton.

Reynolds, A.B. 1996. *Bluebells and Nuclear Energy*. Madison, WI: Cogito.

Rhodes, R., and D. Beller. 2000. "The Need for Nuclear Power." *Foreign Affairs* 79.1: 30–44.

Rowland, R.E. Undated. *Radium Dial Painters: What Happened to Them?* Online at www.rerowland.com/dial_painters.htm.

Royal Society and Royal Academy of Engineering. 1999. *Nuclear Energy: The Future Climate*. London, UK: Royal Society.

Seligman, H. 1990. *Isotopes in Everyday Life*. Vienna: International Atomic Energy Agency.

Shapiro, J. 1981. *Radiation Protection: A Guide for Scientists and Physicians*. 2nd ed. Cambridge, MA: Harvard University Press.

Shcherbak, Y.M. 1996. "Ten Years of the Chernobyl Era." *Scientific American* April: 44–49.

Sims, G. 1981. *A History of the Atomic Energy Control Board*. Ottawa: Canadian Government Publishing Centre.

——. 1990. *The Anti-Nuclear Game*. Ottawa: University of Ottawa Press.

Snell, V.G., and J.Q. Howieson. *Chernobyl: A Canadian Perspective*. Atomic Energy of Canada Limited. Report 910580PA (revised).

Snow, C. P. 1964. *The Two Cultures: And a Second Look*. Cambridge, UK: Cambridge University Press.

——. 1981. *The Physicists*. Boston, MA: Little, Brown.

Stacey, W.M. 2001. *Nuclear Reactor Physics*. New York, NY: Wiley and Sons.

Steane, R. 1997. "Uranium Update." *Canadian Nuclear Society Bulletin* 18.1: 28–31.

Tammemagi, H. 1999. *The Waste Crisis, Landfills, Incinerators, and the Search for a Sustainable Future.* New York, NY: Oxford University Press.

——and N.L. Smith. 1974. "A Radiogeologic Study of the Granites of South-West England." *Journal of the Geological Society* 131: 415–427.

Torgerson, D. F. 2000. "Reducing the Cost of CANDU." *Canadian Nuclear Society Bulletin* 20.4: 22–25.

United Nations. 2005. *Chernobyl's Legacy: Health, Environmental and Socio-Economic Impacts.* United Nations Chernobyl Forum.

UNSCEAR. 1988. *Sources, Effects, and Risks of Ionizing Radiation.* United Nations Scientific Committee on the Effects of Atomic Radiation. New York, NY: United Nations.

——. 1993. "Sources and Effects of Ionizing Radiation." United Nations Scientific Committee on the Effects of Atomic Radiation. New York, NY: United Nations.

US National Committee on Radiological Protection and Measurements. 1980. *Influence of Dose and its Distribution in Time on Dose-Relationships for Low LET Radiation.* NCRP. Report No. 64.

Whitlock, J. 1994. "McMaster Nuclear Reactor Turns 35." *Canadian Nuclear Society Bulletin* 15.1: 2–5.

Zeyher, A. 1997. "Targeted Isotope Homes in on Hodgkin's Disease." *Nuclear News* October: 58–62

Index